T0207204

SpringerBriefs in Molecular Science

More information about this series at http://www.springer.com/series/8898

Kunal Roy · Supratik Kar
Rudra Narayan Das

A Primer on QSAR/QSPR Modeling

Fundamental Concepts

 Springer

Kunal Roy
Department of Pharmaceutical Technology
Jadavpur University
Kolkata
India

Rudra Narayan Das
Department of Pharmaceutical Technology
Jadavpur University
Kolkata
India

Supratik Kar
Department of Pharmaceutical Technology
Jadavpur University
Kolkata
India

ISSN 2191-5407   ISSN 2191-5415 (electronic)
SpringerBriefs in Molecular Science
ISBN 978-3-319-17280-4   ISBN 978-3-319-17281-1 (eBook)
DOI 10.1007/978-3-319-17281-1

Library of Congress Control Number: 2015936155

Springer Cham Heidelberg New York Dordrecht London

© The Author(s) 2015
This work is subject to copyright. All rights are reserved by the Publisher, whether the whole or part of the material is concerned, specifically the rights of translation, reprinting, reuse of illustrations, recitation, broadcasting, reproduction on microfilms or in any other physical way, and transmission or information storage and retrieval, electronic adaptation, computer software, or by similar or dissimilar methodology now known or hereafter developed.
The use of general descriptive names, registered names, trademarks, service marks, etc. in this publication does not imply, even in the absence of a specific statement, that such names are exempt from the relevant protective laws and regulations and therefore free for general use.
The publisher, the authors and the editors are safe to assume that the advice and information in this book are believed to be true and accurate at the date of publication. Neither the publisher nor the authors or the editors give a warranty, express or implied, with respect to the material contained herein or for any errors or omissions that may have been made.

Printed on acid-free paper

Springer International Publishing AG Switzerland is part of Springer Science+Business Media (www.springer.com)

Foreword

In our fast moving society, the use of the computer will become more and more extensive. The "material" things will be more and more supported by the "virtual" situations, which will facilitate our life.

Science is the first actor in this direction, and thus of course, the use of computer methods affect all scientific sectors, within this "internal" dialogue between computer science and the other disciplines, looking for possible applications. The Nobel prize for Chemistry in 2013 recognized the formidable progress in chemistry related to the use of advanced, complex modelling approaches in chemistry.

It is easy to imagine that the so-called QSAR/QSPR methods will increase their role. A few decades ago they were a subject of studies of a very restricted, but active group of scientists, who dedicated their time to anticipate some of the pillars of the QSAR/QSPR field. Today, QSAR/QSPR are mentioned not only by a close group of scientists in their research, but are debated by a growing number of stakeholders looking for the opportunities offered by this field. For instance, industry and regulators discuss the use of the QSAR/QSPR models.

Since the amount of data on chemicals is increasing exponentially, the use of QSAR/QSPR models will become a need, for the purpose to manage the data and extract useful lessons processing the data.

However, all these requirements and expectations have to be supported by a sound theoretical basis and by an updated treatment of the field. For these reasons, I welcome the contribution that is offered by this timely book.

Emilio Benfenati
Istituto di Ricerche Farmacologiche Mario Negri

Preface

Quantitative structure–activity/property relationship (QSAR/QSPR) modelling has been used in medicinal chemistry and computational toxicology for a long time. It offers an in silico tool for the development of predictive models towards various activity and property endpoints of a series of chemicals using the response data that have been determined through experiments and molecular structure information derived computationally or sometimes from experiments. Once developed and validated, such models may be used for prediction of the response endpoint(s) for new and untested chemicals and also for obtaining a mechanistic interpretation of the structure–activity/property relationships. Although these techniques have been successful in many lead optimization and risk assessment problems, their use was previously limited to specific groups of researchers in the chemical sciences. With the easy availability of QSAR-related software tools, QSAR/QSPR modelling is now being exercised by a wider class of researchers; however, some of the users might not have proper background theoretical knowledge in the area. It is desired that QSAR/QSPR users should not depend solely on the available software for model development; *instead*, they should have a basic working knowledge of the theoretical aspects and principles of QSAR/QSPR modelling so that they can develop statistically valid and predictive models which can be meaningfully interpreted.

QSAR/QSPR of the present day is different from what it was during the initial days of its evolution in the form of "Classical QSAR". With the introduction of newer (and higher dimensional) descriptors, the use of sophisticated chemometric tools and rigorous validation strategies and integration with other ligand and structure-based approaches, QSAR/QSPR of the present day is a recognized scientific discipline. QSAR/QSPR is also finding newer applications in diverse fields such as modelling properties/toxicities of nanomaterials, ionic liquids, chemical mixtures, cosmetics, etc., making this an area of potential interest.

In this brief, we aim at introducing the fundamental concepts of QSAR/QSPR modelling in a nutshell to students of Chemical Sciences. The basic concepts seeded into the mind of the students would be a *primer* for the development of their further knowledge in the area through practical modelling exercises and/or additional readings.

Kolkata   Kunal Roy
December 2014   Supratik Kar
                                                                                      Rudra Narayan Das

Contents

Chapter 1
QSAR/QSPR Modeling: Introduction

Abstract Development of predictive quantitative structure–activity relationship (QSAR) models plays a significant role in the design of purpose-specific fine chemicals including pharmaceuticals. Considering the wide application of different types of chemicals in human life, QSAR modeling is a useful tool for prediction of biological activity, physicochemical property, and toxicological responses of untested chemical compounds. Descriptors play a crucial role in the development of any QSAR model since they represent quantitatively the encoded chemical information. They not only help in the derivation of a mathematical correlation between the chemical structure information and the response of interest, but also enable exploration of the mechanistic aspect involved in a biochemical process. QSAR analysis is now widely employed as a rational tool for the prediction and design of chemicals of health benefits, industrial/laboratory process, or household applications.

Keywords Descriptors · Physicochemical · Electronic · Structural · Topological · Quantum chemical

1.1 Introduction

Chemistry plays an important role in defining the behavioral manifestations of chemical compounds. Development of suitable techniques which allow modification of the chemical features of molecules is very useful not only in the field of chemistry but also in other branches of natural sciences. Quantitative structure–activity relationship (QSAR) modeling is one such technique that allows the interdisciplinary exploration of knowledge on compounds covering the aspects of chemistry, physics, biology, and toxicology. It provides a formalism for developing mathematical correlation between the chemical features and the behavioral manifestations of (structurally) similar compounds. The entire technique is defined on the basis of a strong mathematical algorithm, and it provides a reasonable basis for

© The Author(s) 2015
K. Roy et al., *A Primer on QSAR/QSPR Modeling*,
SpringerBriefs in Molecular Science, DOI 10.1007/978-3-319-17281-1_1

establishing a predictive correlation model. Apart from providing a mathematical correlation, QSAR technique also enables the exploration of chemical features encoded within descriptors. Descriptors being the quantitative numbers represent attributes of the chemicals and aid in the establishment of a mathematical correlation. Hence, different types of descriptors play a significant role in the identification as well as analysis of the chemical basis involved in a process under consideration. The descriptors also allow the user to modify or 'fine-tune' the existing chemical behavior into a desired one by suitable changes in chemical structures. Furthermore, such analysis employs chemical information from relatively small number of chemicals in deriving a mathematical correlation while allows the prediction of the same response for a large number of chemicals. This particular characteristic is highly important when dealing with biological (or toxicological) data that involve ethical issues related to animal experiment. The QSAR technique proves to be a valuable alternative method in this perspective and is encouraged for the design and development of biologically active molecules as well as in predictive toxicology analysis. The QSAR formalism is also widely employed to serve different purposes of material science toward the design and development of purpose-specific novel and/or alternative chemicals. It may be very interesting to note that historically the earliest inception for the ideology of QSAR modeling emerged from the simple concept of a correlation between response and chemical nature of molecules which remains the same even today after various developments and nourishments in the QSAR algorithmic basis. Broadly, the two main purposes of QSAR can be identified as the development of a mathematical equation or model and the explanation of the modeled chemical features as encoded in descriptors. Presently, development of predictive QSAR models on various endpoints is proposed by different international authorities as a reliable tool of exploring chemical knowledge following a rational basis [1].

1.2 What Is QSAR/QSPR Modeling?

1.2.1 Definition and Formalism

QSAR modeling on a set of structurally related chemicals refers to the development of a mathematical correlation between a chemical response and quantitative chemical attributes defining the features of the analyzed molecules. Hence, such study attempts to establish a mathematical formalism between the behavior of a chemical, i.e., chemical response and a set of quantitative chemical attributes which may be extracted from the chemical structures using suitable experimental or theoretical means. The naming of the study depends upon the nature of the response (also known as 'endpoint') being modeled giving rise to three major classes, namely quantitative structure–property/activity/toxicity relationship (QSPR/QSAR/ QSTR) studies considering the modeling of physicochemical property, biological activity, and toxicological data, respectively. The nomenclature can also be employed to define some more specific endpoints such as quantitative structure—cytotoxicity relationship to denote modeling of cytotoxicity of chemicals. On the other hand, QSPR, i.e., quantitative structure–property relationship modeling, can be employed to designate all such related techniques as any type of biological and toxicological as well as physicochemical behavior may be considered as the 'property' of a given chemical. However, we shall use the term 'QSAR' to denote all such studies. Since a mathematical relationship is developed, such studies allow the prediction of molecular behavior for new chemicals or even hypothetical molecules. Therefore, the basic formalism of QSAR technique can be mathematically represented as follows:

$$\text{Biological activity} = f(\text{Chemical attributes}) \tag{1.1}$$

The basic ideology for the phrase 'chemical attribute' is to denote the features that define the behavioral manifestation, i.e., response of the analyzed chemical compounds. In other words, the chemical attributes are the fundamental information of the chemicals which control the response under study. Since the aim was to develop a mathematical correlation, these features or attributes are precise quantitative chemical information that might be derived using an experimental analysis or suitable theoretical algorithm that diagnoses chemistry of the molecules. Sometimes, information obtained from both the theoretical as well as experimental basis is employed. It is often observed that the behavioral manifestation of any chemical species can be explained by its physicochemical properties which represent the intrinsic molecular nature such as melting point, boiling point, and surface tension. Hence, the chemical attributes in Eq. (1.1) is often described in terms of the information derived directly from the chemical structure and the physicochemical information usually derived using experimental techniques leading to the following expression [1].

$$\text{Response} = f(\text{chemical structure, physicochemical property}) \tag{1.2}$$

Considering the employment of a series of chemical information in presence/absence of physicochemical features, the QSAR equation for a specific response can be mathematically stated as follows:

$$Y = a_0 + a_1X_1 + a_2X_2 + a_3X_3 + \cdots + a_nX_n \tag{1.3}$$

Since we are talking in terms of a mathematical correlation, such equations are better explained in terms of variables. Here, Y is the dependent variable representing the response being modeled, i.e., activity/property/toxicity while $X_1, X_2,..., X_n$ are the independent variables denoting different structural features or physicochemical properties in the form of numerical quantities or descriptors and $a_1, a_2, ..., a_n$ are the contributions of individual descriptors to the response with a_0 being a constant. Hence, we can see that the physicochemical properties can not only be employed as a dependent or response variable giving a structure–property relationship, i.e., QSPR, but they might also be used as independent or predictor variables. QSAR studies may even employ one response parameter, e.g., activity/toxicity as predictor variable for the modeling of another type of activity/toxicity endpoint. Such studies are named as quantitative activity–activity relationship (QAAR) or quantitative toxicity–toxicity relationship (QTTR) or quantitative property–property relationship (QPPR) modeling, as appropriate. It will be interesting to note that although the modeled response should be quantitative in order to develop a regression model, it might also be categorical entities which may be used for development of classification models. However, the predictor variables in QSAR modeling should always be quantitative.

The QSAR analysis is principally aimed at quantification of chemical information followed by developing a suitable interpretative relationship addressing a given response. The extracted chemical or physicochemical information can be utilized for modification of chemical structures leading to the 'fine-tuning' of the properties and biological response, e.g., decreased lipophilicity, enhanced activity, and reduced toxicological manifestation. Thus, mathematics here serves as a tool for deriving a suitable relationship which is then exploited as per the requirement of the designer [2]. On a much broader perspective, QSAR studies encompasses avenues of chemistry and physics accounting for intrinsic molecular nature, mathematics and statistics for modeling and calculation, and biology to encompass the involved biochemical interaction. Thus, predictive mathematical models are developed exploring the knowledge of chemistry and biology in a rational way to meet the desired need of the chemicals. Different concepts and perspectives of mathematics are tacitly used in order to derive predictive QSAR models which may be used for prediction of endpoint data of a large number of untested chemicals. It might be envisaged that the role of mathematics in QSAR analysis is to provide an abstract backbone for developing a characteristic correlation between chemistry and biology of the investigated chemicals.

The QSAR study can be visualized to comprise of three simple steps, namely (a) data preparation, (b) data processing, and (c) data interpretation for a set of chemicals. The quantitative data are obtained from two major components, namely the response or endpoint to be addressed and the predictor or independent variables (i.e., X variables) defining the chemical attributes. The response data can be activity (e.g., anti-malarial, anti-oxidant, anti-arrhythmic, anti-HIV, and anti-cancer), property (e.g., aqueous solubility, n-octanol/water partition coefficient, melting point, surface tension, critical micelle concentration value, and chromatographic retention), or toxicological (e.g., organ- or disease-specific acute/chronic toxicity outcomes such as carcinogenicity, skin-irritation, genotoxicity, and hepatotoxicity as well as toxicity toward environment in terms of death of specific indicator organisms such as *Tetrahymena*, daphnids, bacteria, fungi, and fish) behavior of chemical compounds. The first step, i.e., the preparation of data involves arrangement and conversion of the data in a suitable form. The response data for various biological and toxicological endpoints are usually obtained in two forms, namely 'dose-fixed response' pattern where the dose or concentration of a chemical required to produce a desired fixed response is measured and 'response-fixed dose' pattern in which the response elicited by a chemical at a fixed dose (concentration) is opted for. An example of the first pattern may be EC_{50} (effective concentration in 50 % population), IC_{50} (concentration required for inhibition of 50 % population), LD_{50} (the dose required to kill half of the total population), etc. Since response values for these analyses being obtained from multiple assays at different dose or concentration levels of chemicals, these (i.e., doses required to elicit a fixed response) are preferably used as the independent variable (Y) in QSAR studies. Hence, a model can be developed from the information of varying concentrations of chemicals required to exhibit a fixed biological (or toxicological) response. One important treatment of the response variable is its logarithmic transformation allowing conversion of a wide range of response data (activity/property/toxicity) into a smaller scale. Another reason for this logarithmic data conversion is that biological/toxicological data give a parabolic curve for the dose–response relationship while the corresponding log dose–response relationship for the same data yields a sigmoidal curve that bears a linear middle portion rendering the modeling easy. It might be noted that the unit for the concentration of chemicals is expressed in molar terms, i.e., M, mM, μM, and nM and hence a chemical eliciting a fixed response at a lower concentration (C) than others actually possesses higher activity or toxicity profile. Hence, activity or toxicity profile of chemicals bears an inverse relationship with their concentration. For all practical purposes, an inverse of the concentration term is usually employed for modeling biological or toxicological data, i.e., $\log 1/C$ or $-\log C$. A data set of chemicals subjected to QSAR analysis is also expected to possess a sufficiently wide range of response data spanning at least 3–4 log units. Furthermore, all the compounds employed for a specific modeling operation are supposed to have a same mechanism of action toward the chosen response. The quantitative data for the predictor variables are obtained from experimental observations usually comprising of different physicochemical measures as well as theoretical calculations. The theoretical computation involves consideration of chemical theories that might be appended with a suitable encoding algorithm. Finally, a data matrix is prepared in which rows present different chemicals in the data set while the response variable and several independent predictor variables are presented in columns. Following the preparation of the data, the modeler needs to process it toward the goal of developing a mathematical equation or model. It may be noted that the data-processing step usually includes several pretreatment operations prior to model development such as removal of inter-correlated features and division of data set which have been discussed in later parts of this book. The data matrix comprising of response and descriptors can be subjected to linear as well as nonlinear model development in combination with a suitable feature selection algorithm. Multiple linear regression (MLR) and partial least squares (PLS) are the representative techniques for the development of linear correlation models while genetic algorithm (GA), stepwise algorithm, etc. can serve methods for variable selection (i.e., feature selection). The nonlinear modeling approaches include artificial neural network (ANN), support vector machine (SVM) and so on. As we can see that the data-processing step including model development involves handling a significant amount of data, such studies should be associated with proper statistical tests. QSAR studies employ computation of several statistical measures and metrics to characterize the quality, stability, and validation of the models. The final operation, i.e., the interpretation of the developed model, is very crucial and it requires a thorough knowledge on the biochemical aspects of the molecules toward the response being modeled. It might be noted that QSAR modeling eventually attempts to establish a chemical basis for specific phenomena such as activity, property, or toxicity by the development of a suitable correlation equation or model. Since, all the chemicals in a data set are assumed to act via same mode of action with respect to a specific response, establishing a mechanistic foundation opens two doors: (a) prediction of the response of existing untested or new chemicals and (b) design and development of completely new chemicals possessing the desired activity/property/toxicity profile. The incorporation of a mathematical algorithm makes the QSAR technique a sound and rational tool [1, 2]. Figure 1.1 presents a simple overview of the QSAR formalism.

Encoding of the chemical features in QSAR analysis is done using a suitable mathematical algorithm. The aim was to perform a definite diagnosis of chemical structural features followed by the derivation of quantitative numbers also known as 'descriptors.' These descriptors carry explicit structural information and are used to establish a correlation with a response of interest. Hence, in a simple terminology, descriptors provide the basis for quantitative depiction of chemical structure, i.e.,

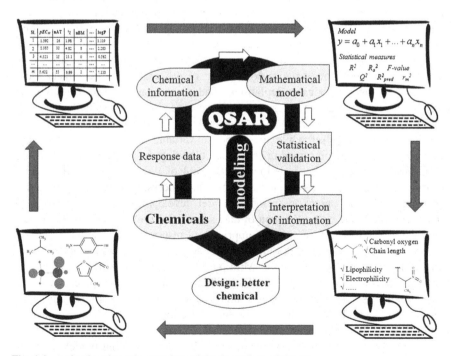

Fig. 1.1 A simple schematic overview of the formalism of QSAR

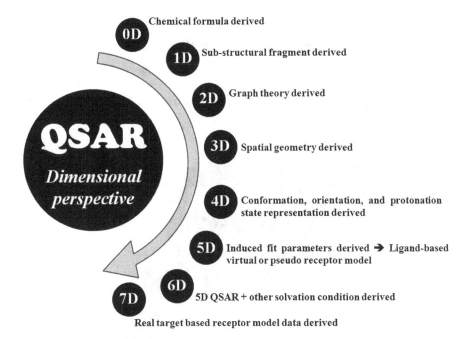

Fig. 1.2 The dimensional perspective of QSAR technique

quantitative numbers derived from a suitable mathematical operation of chemical information. Now, considering the mathematical basis involved in the quantification of chemical information, descriptors can present the dimension of the corresponding QSAR analysis. Since, the extraction of chemical information involves several hypothetical assumptions, QSAR study can be overviewed from a dimensional perspective. In Fig. 1.2, we have outlined QSAR techniques obtained using varying dimensional chemical information. However, based on the mathematical algorithm involved for developing a quantitative correlation, QSAR analysis can be conveniently classified into regression and classification types. The former type of analysis explicitly involves quantitative response values while in case of the classification analysis, one can perform classification of the data into predefined groups or classes. Figure 1.3 shows the mentioned QSAR methods with representative examples in each case.

1.2.2 Objectives of QSAR: Key Features

The principal objective of any QSAR analysis lies in the rational development of a mathematical model accompanied with the exploration of the chemical information involved therein. Such modeling always uses comparatively less amount of data of

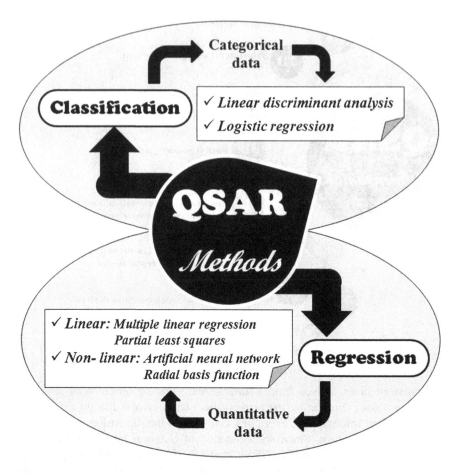

Fig. 1.3 Types of QSAR analysis based on the employed mathematical algorithm for developing correlation

chemical response and allows the prediction for a relatively large number of compounds. This provides an opportunity for this technique to be utilized in various fields. Table 1.1 briefly summarizes the potential key features of the QSAR formalism with an overview of the corresponding applications thereof.

1.2.3 Background

Chemistry serves an essential role for the interdisciplinary exploration of knowledge on the behavioral manifestation of chemicals. Different types of chemicals influence the lives of the human being covering the aspects of industrial use,

Table 1.1 An overview of the key features of the QSAR formalism

Sl. No.	The key objectives and related implications	Brief description
1	Prediction of given response: activity/property/toxicity	A mathematical model is developed with the aim of predicting response of structurally similar chemicals. Usually the prediction is performed for chemicals not included in developing the models. Such chemicals are termed as test set or external set chemicals. Usually a chemical applicability domain is developed using the modeling set (training set) and the prediction of any untested or a new chemical lying within the domain is considered reliable
2	Reduction and replacement of experimental (laboratory) animals	A QSAR study reduces animal experimentation during the preclinical stages of development of drugs since it uses limited chemical response data. The same advantage is also applied in predictive toxicology modeling. Based on the '3R' concept of Russell and Burch, namely replacement, reduction, and refinement of animal experiment in scientific studies, QSAR appears to provide a valuable alternative solution to such ethical issue. Authoritative bodies such as ECVAM, REACH regulation of European Union, office of toxic substances of US-EPA, and OECD propose the use of QSAR as studies alternative to in vivo experiment
3	Virtual screening of library data	Since QSAR leads to the development of an explicit mathematical equation, it can be employed for the screening of chemical library comprising of a large number of compounds. The information derived from descriptors can be utilized as reasonable filtering conditions toward the selection of desired compounds. Examples of some commonly used chemical library are ZINC, DUD benchmark, PubChem, ChemBank, ChEMBL, DrugBank, and Inter-bioscreen.
4	Diagnosis of mechanism	The nature of descriptive information encoded by the descriptors plays a crucial role in this perspective. Establishment of a probable mechanistic interpretation involves defined knowledge on the endpoint especially if it is biological or toxicological. The extracted chemical information is correlated with the corresponding response of interest considering the coefficient of the variables.
5	Categorization of data	A classification algorithm of QSAR allows discrimination of chemicals into groups when the response data are categorical. Such operation is primarily important in the assessment of chemical

(continued)

Table 1.1 (continued)

Sl. No.	The key objectives and related implications	Brief description
		toxicity where categorization of data into different levels of hazard such as high, low, and moderate seem useful
6	Optimization of lead molecules	One of the principle objectives of the QSAR study is the design of purpose-specific chemicals with desired response value. This principle is highly useful during the structural optimization of 'lead' molecules in a drug-designing project in order to get molecules with desired properties. QSAR studies along with other in silico methods can be suitably used toward the successful design and development of drug molecules
7	Structural refinement of synthetic target molecules	It is possible to incorporate the findings of previous QSAR observations during the structural modification process. In a study, Hansch depicted the use of prior knowledge of lipophilicity in eliminating the CNS side effect of the drug Sulmazole [Hansch C. Drug Inf J 1984;18:115–22]

laboratory and institutional applications, as well as household consumption. Hence, it has been a goal of the scientific community to study and search for the information that defines the behavior of the chemicals. The structure–activity relationship emerged as a notion for establishing a link between the chemical structures and their elicited response in a quantitative manner. It will be interesting to note that development of different chemical principles has assisted in the development of QSAR studies. Before going into the details of historical development of the QSAR paradigm, we would like to discuss a few basic dogma of chemistry of compounds. Chemicals are governed by different types of forces and energies that control their physicochemical behavior and the elicited response thereof. Attractive and repulsive forces are the resultant outcomes of intra- and inter-molecular bonding energies of chemicals under the influence of their electronic orbital interactions. All forces must be in an energetically favorable state of balance for the initiation of any kind of molecular interaction. It is to be noted that the attraction (cohesion between similar entities or adhesion between different entities) and repulsion forces might operate simultaneously during a molecular interaction. Considering the biological (and toxicological) response elicited by chemicals, different types of forces play a crucial role for instituting interaction between a chemical and a biomolecule. The physicochemical nature of compounds can be described by three principle phenomena, namely hydrophobic, steric, and electronic effects while various bonding interactions include covalent bond, hydrogen bond, ionic interaction, and dipolar interaction. All these forces and interactions function accordingly when a chemical interferes with a biological system and thereby elicits suitable response. Table 1.2 presents an overview of the mentioned forces and bonding interactions [3, 4].

The name of Mendeleev may be cited as one of the earliest scientists in the field of chemistry who used the concept of chemical correlation in 1870s by formulating the rule of eight. It is believed that the ideology of QSAR emerged in the field of toxicology and later it was supported by experiments in physical organic chemistry. With the progress of time, the concept of chemical correlation became firm in presence of strong mathematical formalisms and with advancements of chemical and physical principles. Cros observed the toxicity of primary alcohols to be correlated with their aqueous solubility in 1863 which implicates that the initial fundamental basis for QSAR modeling was emerged from the toxicological study. However, Crum-Brown and Fraser are considered to be the pioneer in the realm of QSAR modeling who represented physiological action in terms of 'chemical constitution' using a mathematical expression (Eq. 2.1) in 1868, although the phrase 'chemical constitution' was not a well-explained concept at that time.

$$\phi = f(C) \tag{1.4}$$

This observation was followed by Richardson who observed a proportional relationship of narcotic effects of primary alcohols with their molecular weight in 1869. Reynolds and Richet were the next to observe the correlational behavior of chemical nature with corresponding physiological response. Further confidence to the mathematical relationship proposed by Crum-Brown and Fraser was added by Meyer, Overton, and Baum though their work of correlating biological potency of narcotic substances with olive oil/water partition coefficient. It will be noteworthy to mention here that following extended studies, Overton depicted a proportionate mechanistic relationship between increased chain length of the studied compounds and their narcotic behavior in tadpole. Furthermore, he reported different toxicological outcomes of morphine in human and tadpoles and considered a change in the structure of protein in the studied genus. At the beginning of the twentieth century, Traube [5] observed surface tension of chemicals to be related with their narcotic potency which was later modified by Seidell [6] by depicting a similar correlation when solubility and partition coefficient measures were considered [7]. Hence, we can see that the initial observations of QSAR analysis stemmed from toxicological studies and the correlating parameters were principally physico-chemical attributes. In 1933, Ferguson added a thermodynamic basis to it by proposing the narcosis behavior to be linked with the relative saturation of the substance in the applied phase. The next notable development was the exploration of the ionization of chemical species. Albert et al. [8] reported his decisive work on the ionization and shape of aminoacridine compounds in correlating their bacteriostatic potential [7]. This was followed by Bell and Roblin who also performed similar studies on sulfonamides in 1942. The study of developing quantitative descriptors for mathematical correlation models was put into light by Hammett who introduced the decisive electronic substituent constant measure Hammett sigma (σ) in relating the relative reaction rate of *meta*- and *para*-substituted benzoic acid derivatives. This study represents an essential foothold since it was the first

Table 1.2 An overview of the various forces and bonding interactions involved

Type of force/bonding interaction	Brief description
The physicochemical effects	
Hydrophobic effect	Important for eliciting activity/toxicity in biological systems. Also crucial in estimating process efficiency in case of industrial chemicals. Measurement of *n*-octanol/water partition coefficient gives a good measure of this feature in the biological system. Computationally derived measures, viz. CLOGP, MLOGP, AlogP98, etc. can be determined for the whole molecules, while it is also possible to compute hydrophobicity contributions of individual fragments
Electronic effect	This includes different types of dispersion forces, charge transfer complex formation, ionic interaction, inductive effect, hydrogen bonding, polarization effect, acid-base catalytic property, etc. and facilitates interaction with biological receptor systems
Steric effect	Such effects correlate with the spatial arrangement of molecules in the three-dimensional space. These are important in monitoring binding of chemicals to biological receptor cavity
The bonding interactions	
Covalent bond	Such bonding presents the strongest interaction (in the biological system) formed by shared electrons between each of the two participating atoms. In the context of biological receptors, formation of covalent bond between receptor and ligand (chemical) depicts irreversible binding. It has a strength of 50–150 kcal/mol
Ionic bond	This involves electrostatic attraction force between two oppositely charged ions. It bears a strength of 5–10 kcal/mol and hence much less stronger than covalent bond in the biological system. The decrease in strength of the bond is proportional to the squared distance between the participating atoms
Hydrogen bond (H-bond)	This is a weak bonding force with a strength of 2–5 kcal/mol. Such bond formation takes place between a hydrogen atom attached to a strongly electronegative atom and another atom of higher electronegativity. Atoms such as O, N, S, and F can take part in such bond formation in an intra- (same molecule) as well as an inter-molecular (different molecule) fashion. Hydrogen bonds play a very important role in exploring binding of a ligand to its biological receptor. H-bonds stabilize the structure of DNA and hence control its functional characteristics. A H-bond is highly directional at the donor atom (i.e., donating atomic H) and such bonding is controlled by the orbital spatial distribution of the acceptor site (i.e., accepting atomic H) along with a dipolar orientation of the donor group
Hydrophobic force	

(continued)

Table 1.2 (continued)

Type of force/bonding interaction	Brief description
	It measures the dislikeness of molecules toward water. Involvement of such force leads to a favorable change in entropy of a system. Addition of a chemical possessing a nonpolar group in water limits the free Brownian movement of water molecules which is overcome by the reduction of contact of water with the nonpolar interface causing aggregation of the chemical. Addition of a $-CH_2-$ group imparts a strength of 0.37 kcal/mol to a molecule due to hydrophobic force
van der Waals interaction	Such force is relatively weak with a strength varying from 0.5 to 1 kcal/mol and is nonionic in nature. Presence of electronegative atoms such as O and N in molecule cause drawl of the electron cloud toward themselves leading to the formation of a partial positive ($\delta+$) and a partial negative ($\delta-$) charge, i.e., dispersion of charges within the molecule. Two such participating molecules establish a weak bonding interaction. The involved forces are characterized as Keesom force, Debye force, London force, etc. depending on the nature of the established dipole interaction
Pi–pi (π–π) stacking interaction	This is a special type of non-covalent attractive force taking place between unsaturated systems such as arenes. Here, two planar molecules lead to the formation of a stacked geometric complex involving a non-covalent bonding interaction such that the solvent exposed surface area of the complex is minimized. Usually three stacked geometries are identified, viz. parallel-displaced, T-shaped edge-to-face, and eclipsed face-to-face
Charge transfer complex	This is an electron-donor and electron-acceptor complex formed between Lewis bases and Lewis acids. Charge transfer complexes are identified by specific absorption band in the UV-visible range which is different than both the donor and acceptor moieties
Orbital overlapping interaction	Overlapping of pi orbitals leads to the formation of dipole–dipole force of attraction. The π-orbital electron cloud in an aromatic system imparts a negative charge above and below the ring while keeping the equatorial hydrogen atoms positively charged and thereby allowing a dipole–dipole-type interaction between two aromatic systems due to overlap of orbital electron densities. Presence of a lone electron pair containing substituent in aromatic system can influence such bonding
Ion-dipole and ion-induced dipole interaction	Such forces have been found to influence the aqueous solubility of crystalline systems considering water as a dipolar molecule. Here, the cationic and anionic parts of a molecule, respectively, interact with the partially negatively charged oxygen and partially positively charged hydrogen atom of water molecule

established linear free energy relationship (LFER) in the QSAR modeling paradigm as shown by the following equations:

$$\log(k_X/k_H) = \rho \cdot \sigma_X \tag{1.5}$$

$$\log(K_X/K_H) = \rho \cdot \sigma_X \tag{1.6}$$

here k_H and k_X are the rate constant terms for unsubstituted and substituted benzoic acid derivatives, respectively, while K_H and K_X denote their respective equilibrium constants. σ_X represents the Hammett electronic constant of the substituent X, and ρ is the reaction constant term. Since ionization constant terms have been employed to depict σ, the above-mentioned equations are related to the popular free energy equation (see below) and termed as the LFER model.

$$\Delta G^0 = -RT \ln K \tag{1.7}$$

In Eq. (1.7), G denotes Gibbs free energy change, R is the ideal gas constant, and T is the ideal temperature (in Kelvin). Taft devised the first steric descriptor, i.e., the Taft steric parameter E_S in this LFER formalism defining the rates of base- and acid-catalyzed hydrolysis of aliphatic esters and provided an option for separating the effects of polar, steric, and resonance contributions. The next pioneering contribution was the development of Hansch equation in the early 1960s. Corwin Hansch is also credited the title of the 'Father of modern QSAR' who performed studies on plant growth regulators using relative hydrophobicity measure of substituent (π). The linear form of the equation was devised by Fujita and Hansch by the incorporation of Hammett constant term. A general form of the equation is presented below which has undergone several modifications in subsequent times.

$$\log 1/C = k_1\pi + k_2\sigma + k_3 E_s + k_0 \tag{1.8}$$

Here, k_0 represents a constant and k_1, k_2, and k_3 are the coefficient terms of the respective equation variables. Considering the 'random walk process' by drug molecules inside a biological system, Hansch formulated a parabolic relationship which was later extended by incorporating electronic and steric parameters. Both the equations are presented below:

$$\log 1/C = -a(\log P)^2 + b \log P + \text{constant} \tag{1.9}$$

$$\log 1/C = -a(\log P)^2 + b \log P + \rho\sigma + \delta E_s + \text{constant} \tag{1.10}$$

Free and Wilson instituted another approach of QSAR model development on a series of congeneric chemicals by using summed contribution of the parent moiety and structural fragments to represent biological activity.

$$\text{BA} = \sum a_i x_i + \mu \tag{1.11}$$

Here, μ represents the contribution of the parent moiety while a_i denotes the contribution of individual structural fragments with the indicator variable x_i showing their presence ($x_i = 1$) or absence ($x_i = 0$). This equation was later modified by Fujita and Ban who implemented logarithmic activity term to keep the response variable at same level with other free energy terminologies. Considering the scope of this chapter, we will not go into an exhaustive discussion on historical avenues of QSAR modeling. The mentioned discoveries have been pioneering ones and we can see that how the idealism of correlation of 'chemical constituents' with response became a mathematical relationship through the journey involving physicochemical and thermodynamic concepts [7]. Following the development of the LFER model, the mathematical basis for QSAR was well established. Mathematical principles have much more profound impact on theoretical chemistry including QSAR analysis. In the late 1940s, studies on 'chemical graph theory' that involves concepts of mathematics and chemistry led to the development of quantitative descriptors on a purely theoretical basis. Wiener and Platt were the first to develop graph theory-based quantitative topological variables in 1947 known as Wiener index and Platt index, respectively, and reported predictive QSPR models on boiling points of hydrocarbons. This study opened a complete new possibility in the field of theoretical chemistry especially with reference to QSAR formalism that subsequently led to the developments of minimum topological difference (MTD) method of Simon, connectivity index parameters by Randić, Kier and Hall, and many more. This graph theoretic depiction of chemical structures were purely on two-dimensional basis and simultaneous studies on three-dimensional molecular geometry also led to the development of different three-dimensional attributes. Presently, several hundreds to thousands of algorithms are presented to encode molecular features and generate quantitative descriptors employing varying dimensionality, which can be used for QSAR modeling using various statistical methods. However, it will be interesting to note that the sole objective of all such methods and techniques remains the same that initially started with the journey of finding a clue to correlate response with 'chemical constitution' which was a mere composition considered at that time [1, 2, 7]. Figure 1.4 summarizes the pioneering achievements that led to the historical evolution of the QSAR formalism.

1.2.4 Importances of QSAR

Although the development of predictive QSAR/QSPR/QSTR models appears to be a relatively simple task, it has got enormous applications in serving the need of scientific fraternity. It has always been a matter of curiosity that how it is possible for different chemical agents to exert different response profile, and sometimes it is rather astonishing that even the same chemical can elicit different biological actions. Hence, the chemical features appear to be very crucial in determining behavior of chemicals. QSAR techniques can provide several advantages in terms of model predictivity and utilization of limited experimental resources, employing less

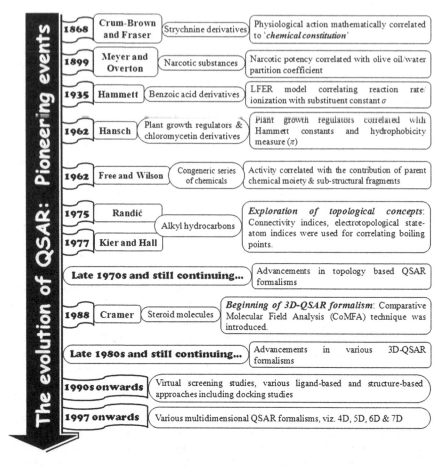

Fig. 1.4 A summary of the pioneering discoveries that led to the gradual evolution of the QSAR study

computational time. Such features encourage the use of QSAR and related techniques in costly research programs such as drug-discovery and development where it can provide valuable information by aiding rational designing strategy with minimal cost involvement. Furthermore, since the QSAR technique can allow the prediction of a chemical response of relatively large number of compounds (within the chemical domain) by using response data of limited number of chemicals, it is widely employed in predictive toxicology analysis for the assessment of chemical hazards. Figure 1.5 depicts an overview of the representative advantages provided by QSAR modeling studies. It may be noted that QSAR helps in achieving efficient, effective, safe, and environmentally benign chemicals and processes thereof and thereby facilitates a 'sustainable chemical' process [2].

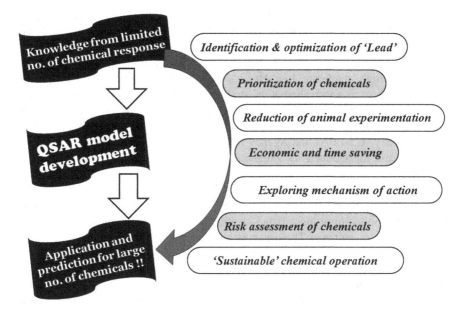

Fig. 1.5 The major advantages obtained from QSAR modeling analysis

1.2.5 QSAR and Regulatory Perspectives

The idealism of developing predictive models using the QSAR techniques is being acknowledged and prescribed by several international regulatory bodies. The following aspects are addressed by different regulatory bodies with the aim of performing risk assessment of chemicals.

1. Assessment of chemical hazard: It comprises identification as well as dose—response characterization of hazard, including classification and labeling of the chemicals.
2. Assessment of exposure.
3. Assessment of hazard and exposure.
4. Identification of persistent, bioaccumulative, and toxic (PBT) as well as very persistent and very bioaccumulative (vPvB) chemicals.

It is obvious that determination of chemical toxicity involves a sound amount of animal experiments in order to generate reliable chemical response data. Hence, it is one of the prime objectives of any hazard assessment strategy to search for suitable alternative method that will reduce animal experimentation. QSAR plays a significant role in this context since it employs comparatively less amount of response data and can predict the same for a large number of chemicals. The QSAR technique complies with the '3R' principle of Russell and Burch, namely replacement, reduction, and refinement of animals in biological experiments and aids in regulatory assessment by performing prioritization of chemicals as well as filling of data gaps. Furthermore, modeling of categorical data (if present) becomes an important aspect here since the toxicological response of chemicals can be categorized into several groups or classes and hence designating different levels of hazards, viz. high, moderate, low, etc. The regulatory agencies which purport the use of QSAR as a valid alternative strategy to animal experiment include the European Centre for the Validation of Alternative Methods (ECVAM) of the European Union, the Office of Toxic Substances of the US Environmental Protection Agency (US-EPA), the Agency for Toxic Substances and Disease Registry (ATSDR), and the Council for International Organizations of Medical Sciences. The European Commission introduced the REACH (Registration, Evaluation, Authorization, and Restriction of Chemicals) regulations in 2006 with an aim of performing systemic evaluation of toxicological hazard of existing as well as new chemicals (imported or produced) and identified QSAR as an alternative method for toxicity testing of animals. The organization of economic cooperation and development (OECD) proposed a set of five point seminal guidelines in 2004 for the proper development and validation of predictive QSAR models by its member countries [9]. With the passage of time, QSAR studies have become an essential part of regulatory assessment on a global perspective, and various countries have developed their own 'expert systems' for determining chemical hazards. Expert systems are the computational applications providing a subject-matter expertise to non-experts by the use of definite logical reasoning. Different expert systems contain models on toxicological endpoints that are prepared and maintained by professional personnel as trusted systems with a suitable user interface such that any unknown or new chemical can easily be tested of its toxicity or categorical-hazard using the existing knowledge-base. Table 1.3 gives a representative overview of some commonly used QSAR expert systems.

1.2.6 Applications of QSAR

Chemicals represent an indispensable part of human necessity considering varying applications spanning from laboratory to industrial processes as well as household usage. QSAR presents a suitable option in the rational monitoring of activity/ property/toxicity of chemicals and hence is useful in a wide variety of applications. Since fine-tuning of the behavioral nature of chemicals gives fruitful results for a significantly large class of chemicals involving pharmaceuticals, agrochemicals, perfumeries, analytical reagents, solvents, surface modifying agents, etc., the application area and possibility of the QSAR technique is enormous. In a global perspective, the chemicals modeled using the QSAR method can be overviewed in three major types, namely chemicals of health benefits (drugs, pharmaceuticals, food ingredients, etc.), chemicals involved in industrial/laboratory processes (solvents, reagents, etc.), and the chemicals posing hazardous outcome [persistent organic pollutants (POPs), toxins, xenobiotics, carcinogents, volatile organic

Table 1.3 A representative overview of some QSAR expert systems

Expert system	Web-address	Expert system	Web-address
Open source systems (free)		*Commercial systems (paid)*	
QSAR TOOLBOX (OECD)	http://www.qsartoolbox.org/	Derek Nexus	http://www.lhasalimited.org/products/derek-nexus.htm
Lazar	http://lazar.in-silico.de/predict	HazardExpert	http://www.compudrug.com/hazardexpertpro
Toxtree	http://ihcp.jrc.ec.europa.eu/our_labs/eurl-ecvam/laboratories-research/predictive_toxicology/qsar_tools/toxtree	The BfR Decision Support System (DSS)	http://www.tandfonline.com/doi/pdf/10.1080/10629360701304014
VEGA	http://www.vega-qsar.eu/	TOPKAT	http://www.sciencedirect.com/science/article/pii/0027510794901252
DEMETRA	http://www.demetra-tox.net/	MCASE and CASE Ultra	http://www.multicase.com/
EPI Suite™	http://www.epa.gov/opptintr/exposure/pubs/episuite.htm	Leadscope	http://www.leadscope.com/
TEST	http://www.epa.gov/nrmrl/std/qsar/qsar.html	TerraQSAR™	http://www.terrabase-inc.com/
OncoLogic™	http://www.epa.gov/oppt/sf/pubs/oncologic.htm	ACD/Percepta	http://www.acdlabs.com/products/percepta/physchem_adme_tox/
		MolCode Toolbox	http://www.molcode.com/
		TIMES	http://oasis-lmc.org/products/software/times.aspx

compounds (VOCs), etc.]. In Fig. 1.6, we have attempted to divide the employment of QSAR application in three broad areas, namely drug designing, material science, and predictive toxicology. Some potential areas of material science which can be addressed by employing predictive QSAR modeling have been depicted in Fig. 1.7, while Fig. 1.8 shows some representative endpoints addressed by the QSAR technique in the sphere of assessing predictive toxicology. It might be interesting to note that apart from modeling biological activity and toxicity endpoints, the drug-designing paradigm involves modeling of ADME which aims to monitor the pharmacokinetic profile of drug candidates prior to its synthesis and thereby enhancing the efficacy of the designed compounds inside biological system.

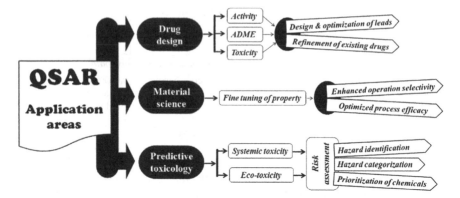

Fig. 1.6 The broad application areas addressed by QSAR modeling studies

Fig. 1.7 Some representative
endpoints addressed by
QSAR analysis in the field of
material science

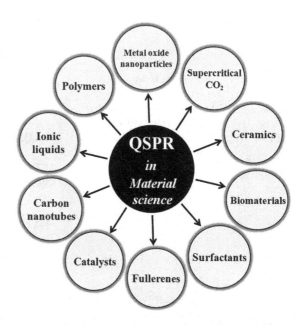

Assessment of toxicity of chemicals principally involves two options, namely assessment of systemic toxicity as well as the monitoring of ecotoxicological hazard. Drugs and pharmaceuticals are capable of posing toxicity to the specific organ system, e.g., hepatotoxicity, cardiovascular toxicity, and nephrotoxicity, while they can also be of serious concern in an environmental perspective since wastewater stream containing even trace amount of such compounds can lead to damage in the ecosystem. Physiologically based pharmacokinetic (PBPK) modeling is another potential area that involves modeling of chemicals such as VOCs using

Fig. 1.8 Some representative endpoints addressed by QSAR study in the realm of predictive toxicology analysis

physicochemical ($\log P$) as well as biochemical parameters (Michaelis constant K_m, maximal velocity V_{max}, hepatic clearance, etc.).

Hence, we can see that the simple ideology of QSAR, i.e., development of a suitable mathematical correlation between the chemical attributes and a response of interest, can be of significant application to serve the human community. Considering the rising health hazard issues and other environmental damage, modern technologies are aimed toward the establishment of a 'sustainable' and 'green' ecosystem that deals with chemical processes that ensure environmental benevolence in terms of efficiency, effectiveness, and safety concerns. QSAR plays an encouraging role in achieving this environmental greenness through the design and development of process-specific chemicals with reduced (or no) hazardous outcomes.

Drug design and development remain the utmost important area addressed by the QSAR formalism. The challenge faced in this perspective is quite higher since the development of a drug molecule is a time consuming as well as costly procedure. Furthermore, the rate of success is also very low since the chance of rejection is very high at any stage of the development paradigm. Figure 1.9 presents an overview of the steps involved in the development of a drug molecule starting from its initial developmental stage. QSAR study can speed up this discovery process by providing rational information on the chemistry of the investigational molecules covering the issues of its contribution to pharmacological behavior, ADME property as well as the toxic outcomes. QSAR can provide valuable information at the stages of design and development and preclinical study, thereby facilitating the outcomes of clinical research and the subsequent approval process. It may be noted that since biomolecular activity involves complex interaction involving

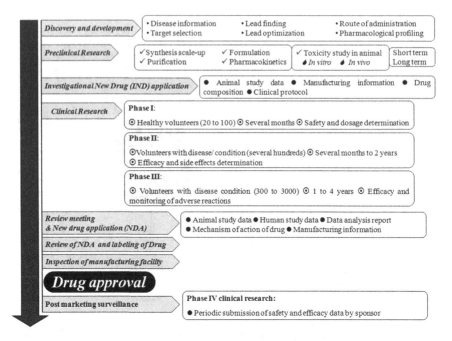

Fig. 1.9 Different phases involved in the development of a drug

Fig. 1.10 Interplay of different in silico techniques with predictive QSAR modeling study

ligand-receptor attributes inside the living system, the development of potential lead molecules would certainly utilize some other techniques as well, namely molecular docking, pharmacophore modeling cheminformatics, and virtual screening along with the QSAR technique. Such techniques are useful in establishing a suitable biochemical correlation for the discovery of drug candidates and can also be applied to other fields as well like toxicophore analysis. Figure 1.10 shows the interplay among various in silico techniques including the QSAR algorithm successfully deployed toward the design of target molecules.

Application of the QSAR technique in combination with other in silico methods has been very fruitful in the drug-discovery paradigm, and some representative examples of such designed drug molecules which were later approved by the US Food and Drug Administration (US-FDA) as drug entities are presented in Fig. 1.11.

Fig. 1.11 Representative examples of drugs designed and developed using different in silico techniques including QSAR modeling analysis. Under individual drugs, the information shown includes the disease indication, the year of US-FDA approval, the proprietary name, and the manufacturing company, respectively

1.3 What Are Descriptors?

1.3.1 Definition

A QSAR model can be expressed as a simple mathematical equation which can correlate the properties (physicochemical/biological/toxicological) of molecules employing diverse computationally or experimentally derived quantitative parameters termed as '*descriptors.*' The descriptors are correlated with the experimental properties (response) using a variety of chemometric tools in order to obtain a statistically significant QSAR model. Molecular descriptors are the '*terms that characterize specific information of a studied molecule.*' They are the '*numerical values associated with the chemical constitution for correlation of chemical structure with various physical properties, chemical reactivity or biological activity.*' The developed equation should provide a significant insight into the essential structural requisites of the molecules which contribute to the biological response of the studied molecules [10]. In other words, the response of a chemical can be mathematically presented as the function of descriptors (Eq. 1.12).

Response (activity/property/toxicity)

$= f$(Molecular information extracted using chemical structure or physicochemical property)

$= f$(Descriptors)

$$(1.12)$$

An ideal descriptor should possess the following features for the construction of a reliable QSAR model:

1. A descriptor should be relevant to a broad class of compounds.
2. A descriptor must be correlated with the studied biological responses while illustrating insignificant correlation with other descriptors.
3. Calculation of the descriptor should be fast and independent of experimental properties.
4. A descriptor should produce different values for structurally dissimilar molecules, even if the structural differences are little.
5. A descriptor should possess physical interpretability to determine the query features for the studied compounds.

A schematic illustration is presented in Fig. 1.12 to depict the steps how a chemical structure is employed to compute descriptors and then utilized in QSAR model development.

1.3.2 Types of Descriptors

Descriptors can be of different types depending on the method of their computation or determination: physicochemical (hydrophobic, steric, or electronic), structural

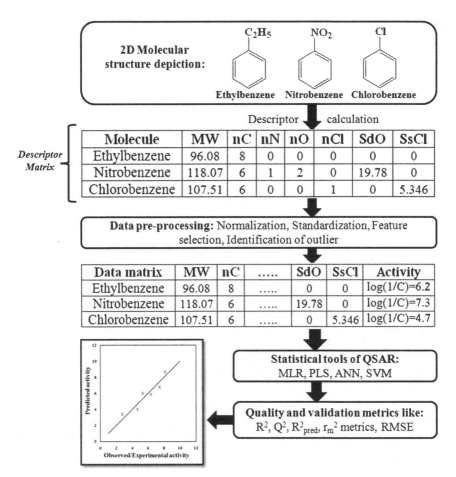

Fig. 1.12 Schematic illustration to show how chemical structures are employed to compute descriptors and QSAR model development

(frequency of occurrence of a substructure), topological, electronic (molecular orbital calculations), geometric (molecular surface area calculation), or simple indicator parameters (dummy variables). In a broader perspective, descriptors (specifically, physicochemical descriptors) can be classified into two major groups: (1) substituent constants and (2) whole molecular descriptors [11, 12]. *Substituent constants* are basically physicochemical descriptors which are designed on the basis of factors, which govern the physicochemical properties of chemical entities. *Whole molecular descriptors* are expansions of the substituent constant approach, but many of them are also derived from experimental approaches.

Descriptors may also be classified based on dimensions. Table 1.4 gives a useful illustration of commonly used molecular descriptors based on dimensions. It is

Table 1.4 Different descriptors employed in the QSAR study based on dimension

Dimension of descriptors	Parameters
0D-descriptors	Constitutional indices, molecular property, atom and bond count
1D-descriptors	Fragment counts, fingerprints
2D-descriptors	Topological parameters, structural parameters, physicochemical parameters including thermodynamic descriptors
3D-descriptors	Electronic parameters, spatial parameters, molecular shape analysis parameters, molecular field analysis parameters and receptor surface analysis parameters

interesting to point out that we have confined our discussion here from 0D- to 3D-descriptors only, though higher dimensional descriptors are also available.

1.3.2.1 2D-Descriptors

Topological

Topological descriptors are calculated based on the graphical representation of molecules and thus they neither require estimation of any physicochemical properties nor need the rigorous calculations involved in the derivation of the quantum chemical descriptors. The structure representation of the molecule depends on its 2D-graphical topology indicating the position of the individual atoms and the bonded connections among them. The formulation of these descriptors is based upon the characterization of chemical structure by graph theory. The graph theoretic determination of the molecular structure involves vertices symbolizing atoms and the covalent bonds representing the edges [13]. In Table 1.5, we have presented the most commonly used topological descriptors along with their formal mathematical definitions briefly, due to their widespread use in QSAR model development.

Structural Parameters

Detailed list of structural descriptors [11] is given in Table 1.6.

Physicochemical Parameters

Physicochemical parameters are designed on the basis of factors, which govern the physical and chemical properties of chemical entities. Due to change in physicochemical properties, absorption, distribution, transport, metabolism, and elimination, behavior of bioactive chemical entities may be changed. The important physicochemical factors affecting bioactivity of drugs and chemical include hydrophobicity, electronic, and steric character of the whole molecules and also the

Table 1.5 A representative overview of topological descriptors used in QSAR model development

Descriptors type	Mathematical definition
Balaban J index	$J = \frac{M}{\mu+1} \sum_{\text{all edges}} (\delta_i \delta_j)^{-0.5}$ where M is the number of edges, μ represents cyclomatic number and δ_i (or δ_j) can be defined as: $\delta_i = \sum_{j=1} \delta_{ij}$
Bond/edge connectivity indices	$\in = \sum_{l=1}^{P_2} [\delta(e_i)\delta(e_j)]_l^{-0.5}$ where $\delta(e)$ corresponds to edge degree and is summed (l) over all the p_2 adjacent edges.
E-state index	$S_i = I_i + \Delta I_i$ where I_i is an intrinsic state parameter and ΔI_i is the perturbation factor. Both the terms are defined as: $I_i = \frac{[(2/N)^2 \delta^v + 1]}{\delta}$ and $\Delta I_i = \sum_{j \neq 1} \frac{(I_i - I_j)}{r_{ij}^2}$ where N is the principal quantum number and r_{ij} being the topological distance between atoms i and j
Extended bond/edge connectivity indices	$^m\in_t = \sum_s \prod_i [\delta(e_i)]_s^{-0.5}$ where m represents the order of the index, t is the type of fragment and $\delta(e_i)$ is the degree of the edge e_i
Extended topochemical atom (ETA) indices	Some basic ETA indices definitions are given below $\alpha = \frac{Z-Z^v}{Z^v} \cdot \frac{1}{PN-1}$, $\beta = \Sigma x\sigma + \Sigma y\pi + \delta$, $\gamma_i = \frac{\alpha_i}{\beta_i}$, $[\eta]_i = \sum_{j \neq i} \left[\frac{\gamma_i \gamma_j}{r_{ij}^2} \right]^{0.5}$, $\varepsilon = -\alpha + 0.3 \times Z^v$, $\psi = \frac{\alpha}{\varepsilon}$ where, α is the core count, β is the valence electron mobile (VEM) count, γ is the VEM vertex count, η is an atom level index, ε is an electronegativity count, and ψ is a measure of hydrogen bonding propensity parameter. Z and Z^v are the respective atomic number and valence electron number; PN corresponds to periodic number; σ and π are the representation of sigma and pi bond, respectively, with their contributions being x and y; δ gives a measure of the resonating lone pair electron in an aromatic system; r_{ij} is the topological distance between two atoms
Kappa shape indices	$^1\kappa = 2\frac{^1P_{\max}^1P_{\min}}{(^1P_i)^2}$; $^2\kappa = 2\frac{^2P_{\max}^2P_{\min}}{(^2P_i)^2}$; $^3\kappa = 2\frac{^3P_{\max}^3P_{\min}}{(^3P_i)^2}$ where, the numbers of one, two, and three path lengths are denoted by ^1P_i, ^2P_i and ^3P_i, respectively. Furthermore, the maximum and minimum path lengths of a specific type may be represented in terms of the number of atoms (A) and thus the corresponding *kappa* shape indices can be defined as follows: $^1P_{\max} = (A(A-1))/2$; $^1P_{\min} = (A-1)$ $^1\kappa = \frac{A(A-1)^2}{(^1P_i)^2}$; $^2\kappa = \frac{(A-1)(A-2)^2}{(^2P_i)^2}$; $^3\kappa = \frac{(A-1)(A-3)^2}{(^3P_i)^2}$ for odd value of A and, $^3\kappa = \frac{(A-2)^2(A-3)}{(^3P_i)^2}$ for even value of A
Kappa modified (alpha) shape indices	The *kappa* indices are modified by using an α term which is defined as: $\alpha_x = \frac{r_x}{r_{C_{sp^3}}} - 1$, where r_x and $r_{C_{sp^3}}$ are the covalent radii of the atom x and sp^3 hybridized carbon atom, respectively. The corresponding alpha-modified *kappa* shape indices are defined

(continued)

Table 1.5 (continued)

Descriptors type	Mathematical definition
	below: $^1\kappa_\alpha = \frac{(A+\alpha)(A+\alpha-1)^2}{(^1P_i+\alpha)^2}$; $^2\kappa_\alpha = \frac{(A+\alpha-1)(A+\alpha-2)^2}{(^2P_i+\alpha)^2}$; $^3\kappa_\alpha = \frac{(A+\alpha-1)(A+\alpha-3)^2}{(^3P_i+\alpha)^2}$ for odd A values and $^3\kappa_\alpha = \frac{(A+\alpha-2)^2(A+\alpha-3)}{(^3P_i+\alpha)^2}$ for even A values
Molecular connectivity index	$^m\chi_t = \sum_{j=1}^{n_m} {}^mS_j$ where, n_m represents the number of t type subgraphs of order m. The term ^mS_j may be defined as follows: $^mS_j = \prod_{i=1}^{m+1} (\delta_i)_j^{-0.5}$ and δ_i for the ith atoms may be defined as: $\delta_i = \sigma_i - h_j$, where, σ_i is the number of valence electrons in σ orbital of the ith atom and h_i represents the number of hydrogen atoms attached to vertex i
Randic branching index (χ)	$\chi = \sum_{\text{all edges}} (\delta_i\delta_j)^{-0.5}$ where, δ_i and δ_j represent the number of other non-hydrogen atoms bonded to atoms (vertices) i and j, respectively, forming an edge ij
Subgraph count index	It is the number of sub-graphs of a given type and order. Subgraph count index is classified from zero order to third order (SC_0, SC_1, SC_2, SC_3). It is notable that third-order sub-graphs are divided into three types on the basis of path, cluster, and ring (SC_3_P, SC_3_C, SC_3_CH)
Valence molecular connectivity index	$^m\chi_t^v = \sum_{j=1}^{n_m} {}^mS_j^v$ Here, the corresponding term δ^v is defined as: $\delta_i^v = \frac{(Z_i^v - h)}{(Z - Z_i^v - 1)}$, where Z and Z^v are the atomic number and the total number of valence electron, respectively, for the ith vertex
Wiener index (W)	$W = \frac{1}{2}\sum_{i=1}^{N}\sum_{j=1}^{N}\delta_{ij}$ where N is the number of vertices or atoms and δ_{ij} is the distance matrix of the shortest possible path between vertices i and j
Zagreb group indices	$\text{Zagreb} = \sum_i \delta_i^2$ where δ_i is the valency of vertex atom i

Table 1.6 Structural parameters used in the development of QSAR models

Parameters	Explanation
Chiral centers	It counts the number of chiral centers (R or S) in a molecule
Molecular weight (MW)	It is the simple molecular weight of a chemical entity
Rotatable bonds (Rotlbonds)	This descriptor counts the number of bonds in the molecule having rotations which are considered to be meaningful for molecular mechanics. All terminal H atoms are ignored
H-bond donor	It counts the number of groups or moieties capable of donating hydrogen bonds
H-bond acceptor	This descriptor calculates the number of hydrogen-bond acceptors present in the molecule

substituents present in the molecules [14–16]. Some formal definitions of physicochemical descriptors commonly used as predictor variables in QSAR analysis are shown in Table 1.7.

Indicator Variables

Indicator variables have been employed in QSAR models due to their simplicity. Substructure-based descriptors can be easily employed as indicator variables. Two sets of compounds which differ from each other only by a substructure existing in one set but not the other can be studied as an entire set when using an indicator variable. The major limitation of this variable is that this approach should only be employed when the two sets of compounds are identical in every respect, except for the substructure being coded with the indicator variable.

Thermodynamic Descriptors

The most commonly used thermodynamic descriptors [11] in QSAR models are described in Table 1.8.

1.3.2.2 3D-Descriptors

Electronic Parameters

Electronic descriptors are defined in terms of atomic charges and are used to describe electronic aspects both of the whole molecule and of particular regions, such as atoms, bonds, and molecular fragments. Electrical charges in the molecule are the driving force of electrostatic interactions, and it is well known that local electron densities or charges play a fundamental role in many chemical reactions and physicochemical properties [11]. The electronic descriptors used in the present studies are summarized in Table 1.9.

Spatial Parameters

Spatial parameters comprise a series of descriptors calculated based on the spatial arrangement of the molecules and the surface occupied by them. The list of spatial descriptors [11] is summarized in Table 1.10.

Table 1.7 Formal definitions of most commonly used physicochemical descriptors in QSAR analysis

Parameter	Definitions
Parameters defining hydrophobic nature	
Partition coefficient	$\log P = \log K_{o/w} = \log \frac{[C]_{n-octanol}}{[C]_{water}}$ where C is the concentration of a solute in the respective mentioned phase (water or *n*-octanol). Usually, compounds having log P value more or less than 1 are considered to be hydrophobic and hydrophilic, respectively.
Hydrophobicity constant (π)	$\pi_X = \log P_X - \log P_H$ where P_X and P_H are the partition coefficient values of the compound with and without specific substituent, respectively. Positive value of π of a given substituent imparts lipophilic character to a molecule and vice versa
Parameters defining electronic nature	
Hammett substituent constant (σ)	$\sigma_X = \log(K_X/K_H)$ where X is a substituent, and K_X and K_H are the equilibrium or dissociation constant with and without the substituent, respectively. Two parameters, namely σ_m and σ_p are widely used representing the respective values for meta and para substituents in an aromatic system
Acid dissociation constant	Acid dissociation constant can be explained by following equation: $K_a = \frac{[A^-][H^+]}{[HA]}$ where A^- is the conjugate base of acid HA and H^+ is the proton. The negative logarithmic function (pK_a) is used for the modeling purpose and can be defined as: $pK_a = -\log_{10} K_a$. It is usually determined using the famous Henderson Hasselbalch equation: $pK_a = pH - \log\frac{[A^-]}{[HA]}$ where, pH is the negative logarithmic concentration of H^+ ion, i.e., $pH = -\log[H^+]$
Parameters defining steric nature	
Taft's steric factor (E_s)	$E_s = \log k_X - \log k_0$ where k_0 and k_X are the rate constants of hydrolysis of an organic compound without having and having substituent X, respectively. The parameter E_s gives a measure of intramolecular steric effect of substituents
Charton's steric parameter (v) and van der Waals radius	Charton found that Taft's steric (E_S) constant is linearly dependent on the van der Waals radius of the substituent, which led to the development of the Charton's steric parameter (v_X). Taft also pointed out that E_s varies parallel to the atom group radius. The Charton's steric parameter can be defined as: $v_X = r_X - r_H = r_X - 1.20$ where, r_X and r_H are the minimum van der Waals radii of the substituent and hydrogen, respectively

(continued)

Table 1.7 (continued)

Parameter	Definitions
Molar refractivity	$MR = \left(\frac{n^2-1}{n^2+2}\right) \times \frac{MW}{\rho}$ where n represents refractive index, molecular weight is denoted by MW, and ρ is the density of the molecule. Molar refractivity provides a measure of volume occupied by an atom or a group
Verloop STERIMOL parameters	Verloop and coworkers developed STERIMOL parameters, which are a set of five descriptors (L, B1, B2, B3, and B4) in order to describe the shape of a substituent. L is the length of the substituent along the axis of the bond between the first atom of the substituent and the parent molecule. The width parameters B1–B4 are all orthogonal to L and form an angle of 90° with each other. The large number of parameters required to define each substituent, and the large number of compounds necessary to incorporate all the parameters into a QSAR, resulted in pruning of the descriptors to L, B1 and B5 with B1 as the smallest and B5 the largest width parameter, which does not have any directional relationship to L
Parachor	An important whole molecular parameter defining the steric nature is parachor which can be explained by following equation $PA = \gamma^{1/4} \cdot \frac{MW}{\rho_L - \rho_V}$ where γ is the surface tension of the liquid, MW is the molecular weight, and ρ_L and ρ_V are the respective densities of the liquid and vapor state. Parachor depends on molecule volume

Table 1.8 Thermodynamic parameters used in the development of QSAR models

Descriptor	Description
AlogP	Log of the partition coefficient using Ghose & Crippen's method
AlogP98	The AlogP98 descriptor is an implementation of the atom-type-based AlogP method
Alogp_atypes	The 120 atom types defined in the calculation of AlogP98 are available as descriptors. Each AlogP98 atom-type value represents the number of atoms of that type in the molecule
Fh2o	Desolvation free energy for water derived from a hydration shell model developed by Hopfinger
Foct	Desolvation free energy for octanol derived from a hydration shell model developed by Hopfinger
Hf	Heat of formation

Table 1.9 Electronic descriptors employed in the construction of QSAR models

Parameters	Explanations
Sum of atomic polarizabilities	It is the summation of atomic polarizabilities (A_i). The polarizabilities are calculated as follows: $$P_a = \sum_i A_i$$ The coefficient, A, is used for calculation of molecular mechanics
Dipole moment (dipole)	This 3D-descriptor represents the strength and orientation behavior of a molecule in an electrostatic field. Both the magnitude and the components (X, Y, Z) of the dipole moment are calculated. It is determined by using partial atomic charges and atomic coordinates
Highest occupied molecular orbital (HOMO) energy	It is the highest energy level in the molecule that contains electrons. When a molecule acts as a Lewis base (an electron-pair donor) in bond formation, the electrons are supplied from this orbital. It measures the nucleophilicity of a molecule
Lowest unoccupied molecular orbital (LUMO) energy	It is the lowest energy level in the molecule that contains no electrons. When a molecule acts as a Lewis acid (an electron-pair acceptor) in bond formation, incoming electron pairs are received in this orbital. It measures the electrophilicity of a molecule
Superdelocalizability (S_r)	It is an index of reactivity in aromatic hydrocarbons, represented as follows: $$S_r = 2 \sum_{j=1}^{m} \left(\frac{c_{jr}^2}{e_j} \right)$$ S_r = superdelocalizability at position r, e_j = bonding energy coefficient in jth molecular orbital (eigenvalue), c = molecular orbital coefficient at position r in the HOMO, m = index of the HOMO The index is based on the idea that early interaction of the molecular orbitals of two reactants may be regarded as a mutual perturbation, so that the relative energies of the two orbitals change together and maintain a similar degree of overlap as the reactants approach one another

Molecular Shape Analysis (MSA) Descriptors

The MSA descriptors are used to determine the molecular shape commonality [11]. Most commonly used MSA descriptors are following: difference volume (DIFFV), common overlap steric volume (COSV), common overlap volume ratio (Fo), non-common overlap steric volume (NCOSV), and root mean square to shape reference (ShapeRMS). A detailed explanation of these MSA descriptors is provided in Chap. 3.

Table 1.10 Spatial parameters used in the development of QSAR models

Parameters	Explanation
Radius of gyration (RadOfGyration)	RadOfGyration is a measure of the size of an object, a surface, or an ensemble of points. It is calculated as the root mean square distance of the objects' parts from either its center of gravity or an axis. This can be calculated as follows: $$\text{RadofGyration} = \sqrt{\left(\sum \frac{\left(x_i^2 + y_i^2 + z_i^2\right)}{N} \right)}$$ here, N is the number of atoms and x, y, z are the atomic coordinates relative to the center of mass
Jurs descriptors	The descriptors combine shape and electronic information to characterize molecules. These descriptors are calculated by mapping atomic partial charges on solvent-accessible surface areas of individual atoms
Shadow indices	These indices help to characterize the shape of the molecules. These are calculated by projecting the molecular surface on three mutually perpendicular planes, i.e., XY, YZ, and XZ. Descriptors depend not only on conformation but also on the orientation of molecule. Molecules are rotated to align principal moments of inertia with X, Y, and Z axes
Molecular surface area (area)	It is a 3D-descriptor that describes the van der Waals area of a molecule. It measures the extent to which a molecule exposes itself to the external environment. It is related to binding, transport, and solubility
Density	This 3D-descriptor is the ratio of molecular weight to molecular volume. This descriptor represents the type of atoms and how tightly they are packed in a molecule. It is related to transport and melt behavior
Principal moment of inertia (PMI)	The moments of inertia are computed for a series of straight lines through the center of mass. These are associated with the principal axes of the ellipsoid. If all three moments are equal, the molecule is considered to be a symmetrical top
Molecular volume (Vm)	This 3D-descriptor is the volume inside the contact surface. It is related to binding and transport

Molecular Field Analysis (MFA) Parameters

The MFA formalism computes probe interaction energies on a rectangular grid around a collection of active molecules. The surface is generated from a 'Shape Field.' The atomic coordinates of the contributing models are used to compute field values on each point of a 3D-grid. MFA evaluates the energy between a probe (H^+ or CH_3) and a molecular model at a series of points defined by a rectangular grid. Fields of molecules are represented using grids in MFA and each energy associated with an MFA grid point can serve as input for the calculation of a QSAR [17].

Table 1.11 List of software tools and online platforms for computation of molecular descriptors

Software/online platform	Weblink
Cerius^2	http://accelrys.com/
CODESSA PRO	http://www.codessa-pro.com/index.htm
Discovery studio	http://accelrys.com/
DRAGON	http://www.talete.mi.it/products/dragon_description.htm
E-Dragon at VCCLAB	http://www.vcclab.org/lab/edragon/
GRID	http://www.moldiscovery.com/soft_grid.php
JME Molecular Editor	http://www.molinspiration.com/jme/index.html
Linux4Chemistry	http://www.linux4chemistry.info/
MOE	http://www.chemcomp.com/software.htm
MOLCONN-Z	http://www.edusoft-lc.com/molconn/
MOLE db	http://michem.disat.unimib.it/mole_db/
MOLGEN-QSPR	http://www.molgen.de/?src=documents/molgenqspr.html
OCHEM	https://ochem.eu/home/show.do
OpenBabel	http://openbabel.org/
PaDEL-Descriptor	http://padel.nus.edu.sg/software/padeldescriptor/
PCLIENT	http://www.vcclab.org/lab/pclient/
QSARModel	http://www.molcode.com/
QuaSAR	http://www.chemcomp.com/feature/qsar.htm
SYBYL-X	http://tripos.com/index.php?family=modules,SimplePage&page=SYBYL-X
Tsar^{TM}	http://www.accelrys.com/products/tsar/tsar.html
Unscrambler X	http://www.camo.com/rt/Products/Unscrambler/unscrambler.html
V-Life MDS	http://www.vlifesciences.com/products/VLifeMDS/Product_VLifeMDS.php

Receptor Surface Analysis (RSA) Parameters

The energies of interaction between the receptor surface model and each molecular model can be used as descriptors for generating QSARs [17]. The surface points that organize as triangle meshes in the construction of the RSA store these properties as associated scalar values. Receptor surface models provide compact, quantitative descriptors, which capture three-dimensional information of interaction energies in terms of steric and electrostatic fields at each surface point. A detailed explanation of these RSA descriptors is provided in Chap. 3.

1.3.3 Software Tools and Online Platforms

QSAR is gaining popularity among the researchers with the development of new and advanced software tools and online platforms which allow them to determine the molecular structural features responsible for compounds activity/property/toxicity. Table 1.11 shows a representative list of most commonly employed software tools and online platforms for the generation of descriptors from molecular structures.

1.4 Conclusion

Development of techniques to fine-tune and modify the chemistry of compounds provides an enormous opportunity toward the development of purpose-specific chemicals. The search for the answer to the query how different chemicals elicit different responses and even the same chemical shows varying behavioral features has led to the exploration of different chemical attributes. Predictive QSAR modeling technique provides an option for developing a mathematical basis for the elicited chemical responses. Since the generated basis is highly rational on the ground of chemical information, such techniques are widely employed to maneuver the needs of the industry as well as academia. The drug-discovery paradigm involving costly and time-consuming steps can be easily rationalized and put under suitable basis using QSAR and other suitable in silico techniques in the preclinical research programmes. The QSAR technique also enables optimization of chemical operations by enhancing the selectivity of various process chemicals. Furthermore, the QSAR technique has profound applications in the risk assessment paradigm considering the minimal engagement of ethical issues related to animal experiment while using regression or classification-based predictive mathematical models. Among other components, descriptors present one of the crucial elements of the QSAR formalism. The ultimate diagnosis of chemical features is preserved in the form of quantitative numbers as descriptors which enables the identification of mechanism of action of a given biochemical process and any modification thereof. It should be noted that 'no' single descriptor can provide any universal solution to chemical problems. Sometimes, the nature of endpoints becomes the determining parameter in choosing suitable descriptors. Although a wide variety of descriptors are available for use, the goal of a modeler should be toward the use or development of descriptors which are easily computable giving an explicit amount of chemical information. Hence, one of the major goals of the modeler should not only be directed toward the development of a good mathematical correlation between response and descriptors, but it should also provide a suitable explanation of the result, i.e., a mechanistic overview such that the QSAR formalism can be used as a rational chemical designing tool instead a 'black-box' method of deriving a mathematical correlation involving a series of abstract mathematical algorithms.

References

1. Todeschini R, Consonni V, Gramatica P (2009) Chemometrics in QSAR. In: Brown S, Tauler R, Walczak R (eds) Comprehensive chemometrics, vol 4. Elsevier, Oxford, pp 129–172
2. Tute MS (1990) History and objectives of quantitative drug design. In: Hansch C, Sammes PG, Taylor JB (eds) Comprehensive medicinal chemistry, vol 4. Pergamon Press, Oxford, pp 1–31
3. Sinko PJ (ed) (2011) Martin's physical pharmacy and pharmaceutical sciences, 6th edn. Lippincott Williams & Wilkins, Baltimore
4. Daniels TC, Jorgensen EC (1982) Physicochemical properties in relation to biological action. In: Doerge RF (ed) Wilson and Gisvold's textbook of organic medicinal and pharmaceutical chemistry, 8th edn. J.B. Lippincott Co., Pennsylvania
5. Traube J (1904) Theorie der Osmose and Narkose. Pflüg Arch Physiol 105:541–558
6. Seidell A (1912) A new bromine method for the determination of thymol, salicylates, and similar compounds. Am Chem J 47:508–526
7. Selassie CD (2003) History of quantitative structure-activity relationships. In: Abraham DJ (ed) Burger's medicinal chemistry and drug discovery, vol 1., Drug DiscoveryWiley, New York, pp 1–48
8. Albert A, Rubbo SD, Goldacre R (1941) Correlation of basicity and antiseptic action in an acridine series. Nature 147:332–333
9. Fjodorova N, Novich M, Vrachko M, Smirnov V, Kharchevnikova N, Zholdakova Z, Novikov S, Skvortsova N, Filimonov D, Poroikov V, Benfenati E (2008) Directions in QSAR modeling for regulatory uses in OECD member countries, EU and in Russia. J Environ Sci Health Part C Environ Carcinog Ecotoxicol Rev 26:201–236
10. Guha R, Willighagen E (2012) A survey of quantitative descriptions of molecular structure. Curr Top Med Chem 12:1946–1956
11. Todeschini R, Consonni V (2000) Handbook of molecular descriptors. Wiley-VCH, Weinheim
12. Livingstone DJ (2000) The characterization of chemical structures using molecular properties. A survey. J Chem Inf Comput Sci 40:195–209
13. Roy K, Das RN (2014) A review on principles, theory and practices of 2D-QSAR. Current Drug Metabol 15:346–379
14. Taylor PJ (1991) Quantitative drug design. the rational design, mechanistic study and therapeutic applications of chemical compounds. In: Hansch C, Sammes PG, Taylor JB (eds) Comprehensive medicinal chemistry, vol 4. Pergamon Press, Oxford; pp 241–294
15. Rekker R (1977) The hydrophobic fragmental constant. Elsevier, Amsterdam
16. Hansch C, Leo A, Hoekman D (1995) Exploring QSAR vol 2: hydrophobic, electronic and steric constants. ACS, Washington DC
17. Hopfinger AJ, Tokarsi JS (1997) In: Charifson PS (ed) Practical applications of computer-aided drug design. Marcel Dekker, New York, pp 105–164

Chapter 2
Statistical Methods in QSAR/QSPR

Abstract QSAR/QSPR studies are aimed at developing correlation models using a response of chemicals (activity/property) and chemical information data in a statistical approach. The regression- and classification-based strategies are employed to serve the purpose of developing models for quantitative and graded response data, respectively. In addition to the conventional methods, various machine learning tools are also useful for QSAR/QSPR modeling analysis especially for studies involving high-dimensional and complex chemical information data bearing a nonlinear relationship with the response under consideration.

Keywords Applicability domain · Chemometric tools · Classification · MLR · Model development · OECD · Validation

2.1 Introduction

QSAR/QSPR models represent mathematical equations correlating the response of chemicals (activity/property) with their structural and physicochemical information in the form of numerical quantities, i.e., descriptors. Suitable statistical methods are deployed to derive a robust mathematical correlation involving small to large number of variables. Various regression- and classification-based methods are used for this purpose. Regression-based approaches are employed when the response data of chemicals are entirely numerical, i.e., quantitative, while qualitative or semi-quantitative chemical response(s) are modeled using classification techniques. It may be noted that the descriptors in both the cases of regression- and classification-based methods will be explicitly quantitative values. The regression-based methods enable the quantitative prediction of the response (activity/property), while classification methods allow categorization of the data points into several groups or classes such as highly active and less active. In addition to the conventional

© The Author(s) 2015
K. Roy et al., *A Primer on QSAR/QSPR Modeling*,
SpringerBriefs in Molecular Science, DOI 10.1007/978-3-319-17281-1_2

methods, machine learning-based methods are also useful in developing QSAR/
QSPR models. It may be noted that the machine learning tools employing artificial
intelligence can also be used to solve regression- and classification-based problems.
Now, apart from the model development formalisms, various statistical tools are
also useful for feature selection from a large matrix of descriptor data. The feature
selection tools enable the use of suitable and relevant descriptors for a particular
response, thereby removing noises from the analysis. Furthermore, the descriptor
data matrix can also be subjected to various pruning methods to reduce intercorrelated and redundant chemical information. The developed QSAR models are also
subjected to several validation tests to check for the reliability of the developed
correlation models. After its development, a QSAR model is usually verified by
employing multiple statistical validation strategies giving an estimation of its predictivity and stability. According to the OECD guidelines, the development of a
QSAR model should comply with unambiguous algorithm strategies and the model
should pass various testes model fitness, robustness, and predictivity. The present
chapter gives an account of various statistical tools used for the data pretreatment,
feature selection, model development, and validation of QSAR/QSPR models.

2.2 Chemometric Tools

Chemometrics is the chemical discipline that uses statistical methods to design
optimal procedures, experiments, and objects, and to provide maximum chemical
information by analyzing chemical data.

2.2.1 Various Chemometric Tools Used in QSAR/QSPR

QSAR/QSPR is basically a statistical approach correlating the response property or
activity data with descriptors encoding chemical information. Such correlation may
be derived either in a regression-based approach (in cases where the response
property is quantitative and available in a continuous scale) or a classification-based
approach (in cases where the response property is graded or semi-quantitative).

The most commonly used regression-based approaches are as follows:

- Multiple linear regression (MLR)
- Partial least squares (PLS)

Some of the common classification-based approaches are as follows:

- Linear discriminant analysis (LDA)
- Cluster analysis

Machine learning tools such as artificial neural network, support vector machine are also very effective in developing predictive models, particularly handling with high-dimensional and complex chemical information data showing a nonlinear relationship with the response(s) of the chemicals. Some of the more popular and commonly used chemometric tools will be briefly discussed in this chapter. However, before any statistical model building method is applied, the QSAR/QSPR data table may be required to be pretreated followed by application of a suitable feature selection method.

2.2.2 Pretreatment of the Data Table

While preparing a QSAR table, care should be taken to ensure that the molecular structures have been correctly drawn or imported, the biological activity (or other response) data have been taken from an authentic source (and they have permissible experimental errors) and the descriptor values have been computed using a validated software. The response data for a QSAR modeling set should ideally have a normal distribution pattern. While clubbing two or more data sets, care must be taken to ensure that all experiments performed to determine the response values have used the same protocol. Care should also be taken to avoid duplicates in the data set. The correct tautomeric form of the structure of the compounds should also be considered. For computation of 3D descriptors, appropriate structure optimization should have been carried out.

When a large number of descriptors have been calculated, an appropriate method to remove less important or redundant descriptors should be applied. One can omit the descriptors with a constant value for all observations and the descriptors showing a very low variance. Only one descriptor among those showing high mutual intercorrelation should be retained. Descriptors showing a very low correlation with the response may also be omitted in order to thin the descriptor pool. In some cases, a suitable scaling of the descriptors may also be required.

2.2.3 Feature Selection

The selection of appropriate descriptors for model development from a pool of a large number of descriptors is an important step in QSAR modeling. Such selection may be done in a variety of ways, including stepwise selection (using a suitable stepping criterion, e.g., 'F-for-inclusion' and 'F-for-exclusion' based on partial F-statistic), all possible subset selection, genetic method, and factor analysis.

2.2.4 Multiple Linear Regression

Multiple linear regression or MLR [1] is a commonly used method in QSAR due to its simplicity, transparency, reproducibility, and easy interpretability. The generalized expression of an MLR equation will be like the following:

$$Y = a_0 + a_1 \times X_1 + a_2 \times X_2 + a_3 \times X_3 + \cdots + a_n \times X_n \qquad (2.1)$$

In the above expression, Y is the response or dependent variable, $X_1, X_2, \ldots, X_n$ are descriptors (features or independent variables) present in the model with the corresponding regression coefficients $a_1, a_2, \ldots, a_n$, respectively, and a_0 is the constant term of the model. The interpretation of contribution of individual descriptors $X_1, X_2, \ldots, X_n$ is straightforward depending on the corresponding coefficient value and its algebraic sign. Each regression coefficient should be significant at $p < 0.05$ which can be checked from a 't' test. The descriptors present in an MLR model should not be much intercorrelated. For a statistically reliable model, the number of observations and number of descriptors should bear a ration of at least 5:1. A MLR model that fits well the given data will lead to a scatter plot (observed vs. calculated) showing a minimum deviation of the points from the line of fit (Fig. 2.1). The quality of a MLR model is determined from a number of metrics as described below.

Fig. 2.1 A scatter plot of the observed and calculated activity for an MLR model

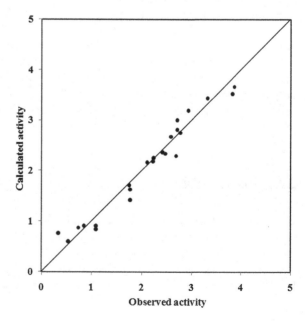

1. Determination coefficient (R^2)
   One can define the determination coefficient (R^2) in the following manner:

$$R^2 = 1 - \frac{\sum (Y_{obs} - Y_{calc})^2}{\sum (Y_{obs} - \overline{Y_{obs}})^2} \tag{2.2}$$

   In the above equation, Y_{obs} stands for the observed response value, while Y_{calc} is the model-derived calculated response and $\overline{Y_{obs}}$ is the average of the observed response values. For the ideal model, the sum of squared residuals being 0, the value of R^2 is 1. As the value of R^2 deviates from 1, the fitting quality of the model deteriorates. The square root or R^2 is the multiple correlation coefficient (R).

2. Adjusted R^2 (R_a^2)
   If one goes on increasing the number of descriptors in a model for a fixed number of observations, R^2 values will always increase, but this will lead to a decrease in the degree of freedom and low statistical reliability. Thus, a high value of R^2 is not necessarily as indication of a good statistical model that fits well the available data. To reflect the explained variance (the fraction of the data variance explained by the model) in a better way, adjusted R^2 which has been defined in the following manner:

$$R_a^2 = \frac{(N - 1) \times R^2 - p}{N - 1 - p} \tag{2.3}$$

   In the above expression, p is the number of predictor variables used in the model development.

3. Variance ratio (F)
   To judge the overall significance of the regression coefficients, the variance ratio (the ration of regression mean square to deviations mean square) can be defined as follows:

$$F = \frac{\frac{\sum (Y_{calc} - \bar{Y})^2}{p}}{\frac{\sum (Y_{obs} - Y_{calc})^2}{N - p - 1}} \tag{2.4}$$

   The F value has two degrees of freedom: $p, N - p - 1$. The computed F value of a model should be significant at $p < 0.05$. For overall significance of the regression coefficients, the F value should be high.

4. Standard error of estimate (s)
   For a good model, the standard error of estimate of Y should be low and this is defined as follows:

$$s = \sqrt{\frac{(Y_{obs} - Y_{calc})^2}{N - p - 1}} \tag{2.5}$$

   It has a degree of freedom of $N - p - 1$.

Note that development of MLR models and computation of various statistical metrics can be done by the use of an open access tool available at http://dtclab. webs.com/software-tools and http://teqip.jdvu.ac.in/QSAR_Tools/ and also from the site http://aptsoftware.co.in/DTCMLRWeb/index.jsp.

2.2.5 Partial Least Squares (PLS)

While handling a large number of intercorrelated and noisy descriptors for a limited number of data points, PLS is a better choice over MLR. PLS, being a generalization of MLR [2], tries to extract the latent variables (LV), which are functions of the original variables, accounting for as much of the underlying factor variation as possible while modeling the responses. Before the analysis, the X- and Y-variables are often transformed to make their distributions fairly symmetrical. The response variables are usually logarithmically transformed and the X variables should be scaled appropriately. The linear PLS finds a few new variables (latent variables), which are linear combinations of the original variables. When the number of LVs is equal to the number of variables, the PLS model becomes same as the MLR model. A strict test of the predictive significance of each PLS component is necessary, and then stopping addition of new components when components start to be nonsignificant. Cross-validation (CV) is a practical and reliable way to test this predictive significance. A PLS equation can be expressed in the same form as in MLR; thus contributions of individual descriptors to the response can be easily found out.

2.2.6 Linear Discriminant Analysis

LDA [3] can separate two or more classes of objects and can thus be used for classification problems. LDA performs the same task as MLR by predicting an outcome when the response property has graded values and molecular descriptors are continuous variables. LDA explicitly attempts to model the difference between the classes of data. In a two-group situation, the predicted membership is calculated by computing a discriminant function (DF) score for each case (Fig. 2.2). Then, cases with DF values smaller than the cutoff value are classified as belonging to one group, while those with values larger are classified into the other group. The DF may take the following form:

$$DF = c_1 \times X_1 + c_2 \times X_2 + \cdots + c_m \times X_m + a \qquad (2.6)$$

where DF is the discriminate function, which is a linear combination (sum) of the discriminating variables, c is the discriminant coefficient or weight for that variable, X is respondent's score for that variable, a is a constant, m is the number of predictor variables. The c's are unstandardized discriminant coefficients analogous

Fig. 2.2 Distribution of compounds in two groups using a discrimination function DF in a LDA analysis

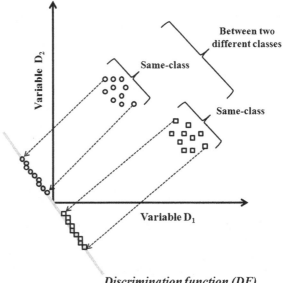

Discrimination function (DF)

to the *beta* coefficients in the regression equation. These c's maximize the distance between the means of the criterion (dependent) variable. Good predictors tend to have large standardized coefficients. After using an existing set of data to calculate the DF and classify cases, any new cases (test samples) can then be classified.

In a stepwise DF analysis, the model is built step-by-step. Specifically, at each step all variables are reviewed and evaluated to determine which one will contribute most to the discrimination between groups. That variable will then be included in the model, and the process starts again.

2.2.7 Cluster Analysis

Unlike LDA, cluster analysis [4] requires no prior knowledge about which elements belong to which clusters. The clusters are defined through an analysis of the data. Cluster analysis maximizes the similarity of cases within each cluster while maximizing the dissimilarity between groups that are initially unknown.

The hierarchical cluster analysis finds relatively homogeneous clusters of cases based on dissimilarities or distances among objects. The most straightforward and generally accepted way of computing distances between objects in a multi-dimensional space is to compute the Euclidean distances or the squared Euclidean distance. It starts with each case as a separate cluster and then combines the clusters sequentially, reducing the number of clusters at each step until only one cluster is left. A hierarchical tree diagram or dendrogram (Fig. 2.3) can be generated to show the linkage points: the clusters are linked at increasing levels of dissimilarity.

Fig. 2.3 Example of a
dendrogram

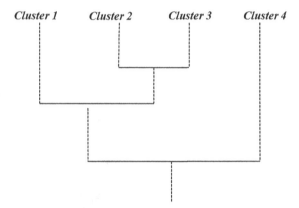

The k-means clustering is a non-hierarchical method of clustering which can be used when the number of clusters present in the objects or cases is known. It is an unsupervised method of centroid-based clustering. In general, the k-means method will produce the exact k different clusters. The method defines k centroids, one for each cluster, placed as much as possible far away from each other. The next step is to take each point belonging to a given data set and associate it to the nearest centroid. When no point is pending, the positions of the k centroids are recalculated. This procedure is repeated until the centroids no longer move.

2.3 Quality Metrics

2.3.1 Importance of Metrics for Determination of Quality of QSAR Models

Advancement in fast and economical computational resources makes it feasible to compute a large number of descriptors using various software tools. As a consequence, one cannot deny the risk of chance correlations with the increasing number of variables included in the QSAR model as compared to the limited number of compounds usually employed for the model development [5]. On the other hand, employing miscellaneous optimization tools, it is feasible to get models that can fit well the experimental data but there always remains a chance of overfitting. Fitting of data does not corroborate a good predictability of the model as the former is a parameter for the statistical quality of the model. This is the main reason why validation tools must be applied on the developed QSAR model to check its predictivity for new untested molecules. A flowchart for the method of development of a dependable QSAR model along with the various validation methods with the metrics commonly used are demonstrated in Fig. 2.4.

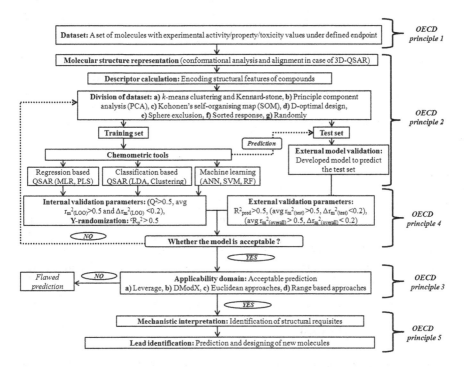

Fig. 2.4 Fundamental steps for the generation of a QSAR model and employed validation methods

2.3.2 Types of Validation

2.3.2.1 The OECD Principles

The OECD principles are the best possible outline of the essential points to be addressed while developing reliable and reproducible QSAR models [6]. The principles were formulated by QSAR experts in a meeting held in Setúbal, Portugal, in March 2002 as the guidelines for the validation of QSAR models, in particular for regulatory purposes. These principles were later approved by the OECD member countries, QSAR and regulatory communities at the 37th Joint Meeting of the Chemicals Committee and Working Party on Chemicals, Pesticides and Biotechnology in November 2004. The five guidelines adopted by the OECD denoting validity of QSAR model are as follows:

- Principle 1—A defined endpoint
- Principle 2—An unambiguous algorithm
- Principle 3—A defined domain of applicability

- Principle 4—Appropriate measures of goodness-of-fit, robustness and predictivity
- Principle 5—A mechanistic interpretation, if possible.

The present challenge in the process of development of a QSAR model is no longer in developing a model that is statistically sound to predict the activity within the training set, but in developing a model with the capability to accurately predict the activity of new chemicals.

2.3.2.2 Internal Validation

Internal validation of a QSAR model is performed based on the molecules used in the model development. It involves activity prediction of the studied molecules followed by estimation of parameters for detecting the precision of predictions. To judge the quality and goodness-of-fit of the model, internal validation is an ideal technique. But, the major disadvantage of this approach is the lack of predictability of the model when it is applied to a new data set [7].

2.3.2.3 External Validation

One cannot judge the predictability of the developed model from internal validation for an entirely new set of compounds, as internal validation considers the chemicals belonging to the same set of compounds used for model development. Thus for external validation, the available data set is usually divided into training and test sets, then subsequently a model is developed with the training set, and then the constructed model is employed to check the external validation employing the test set molecules which are not utilized in the model development process. The external validation ensures the predictability and applicability of the developed QSAR model for the prediction of untested molecules [8].

Selection of Training and Test Sets

In general, the division of the data set into training and test sets must be executed in such a manner that points representing both training and test sets are dispersed within the entire descriptor space occupied by the whole data set and each point of the test set is near to at least one compound of the training set. The following approaches are mostly employed by the QSAR practitioners for the selection of the training and test sets [8]:

1. **Random selection**: The data set may be divided by a mere random selection process.
2. **Based on Y-response**: This approach is based on the activity (Y-response) sampling. The complete range of the response is divided into bins and

compounds belonging to each bin are assigned to the training or test sets randomly or in customized way.

3. ***Based on X-response***: Properties and structural similarity of the molecules are considered for the grouping of similar compounds. After that, a predecided fraction of compounds is assigned to the training or test set manually or in some regular way.

Most commonly employed tools for the rational division of the data sets are:

- k-Means clustering,
- Kohonen's self-organizing map selection,
- statistical molecular design,
- Kennard–Stone selection,
- sphere exclusion, and
- extrapolation-oriented test set selection.

Note that the division of a data set using some common algorithms can be easily done by the use of an open access tool available at http://teqip.jdvu.ac.in/QSAR_Tools/.

Applicability Domain (AD)

1. Concept of the AD
   The AD is defined as a theoretical region in the chemical space constructed by both the model descriptors and modeled response. The applicability domain plays a crucial role for estimating the uncertainty in the prediction of a particular compound based on how similar it is to the compounds employed to construct the QSAR model. Therefore, the prediction of a modeled response using QSAR is applicable only if the compound being predicted falls within the AD of the model as it is unfeasible to predict the whole universe of compounds using a single QSAR model [9].

2. Types of the AD approaches
   The most commonly employed techniques for estimating interpolation regions in a multivariate space are as follows:

   (a) Ranges in the descriptor space,
   (b) geometrical methods,
   (c) distance-based methods,
   (d) probability density distribution, and
   (e) range of the response variable.

The first four approaches are based on the methodology used for interpolation space characterization in the model descriptor space. On the contrary, the last one depends solely on response space of the training set molecules. A compound can be identified out of the AD, if: (a) at least one descriptor is out of range for the ranges approach and (b) the distance between the chemical and the center of the training data set exceeds the threshold for distance approaches. The threshold for all kinds of distance methods is the largest distance between the training set data points and the center of the training data set.

2.3.3 Validation Metrics for Regression-Based QSAR Models

2.3.3.1 Metrics for Internal Validation

The most commonly employed internal metrics are discussed below [10]:

1. Leave-one-out (LOO) cross-validation
   To determine the LOO cross-validation, the training set is primarily modified by eliminating one compound from the set. The QSAR model is then rebuilt based on the remaining molecules of the training set using the descriptor combination originally selected, and the activity of the deleted compound is computed based on the resulting QSAR equation. This cycle is repeated until all the molecules of the training set have been deleted once, and the predicted activity data obtained for all the training set compounds are used for the calculation of various internal validation parameters. Finally, the model predictivity is judged using the pre-dicted residual sum of squares (PRESS) and cross-validated R^2 (Q^2) for the model while the value of standard deviation of error of prediction (SDEP) is calculated from PRESS.

$$\text{PRESS} = \sum \left(Y_{\text{obs}} - Y_{\text{pred}}\right)^2 \tag{2.7}$$

$$\text{SDEP} = \sqrt{\frac{\text{PRESS}}{n}} \tag{2.8}$$

$$Q^2 = 1 - \frac{\sum \left(Y_{\text{obs(train)}} - Y_{\text{pred(train)}}\right)^2}{\sum \left(Y_{\text{obs(train)}} - \overline{Y}_{\text{training}}\right)^2} = 1 - \frac{\text{PRESS}}{\sum \left(Y_{\text{obs(train)}} - \overline{Y}_{\text{training}}\right)^2} \tag{2.9}$$

   In Eqs. (2.7)–(2.9), Y_{obs} and Y_{pred} correspond to the observed and LOO-predicted activity values, n refers to the number of observations, $Y_{\text{obs(train)}}$ is the observed activity, $Y_{\text{pred(train)}}$ is the predicted activity of the training set molecules based on the LOO technique. The threshold value of Q^2 is 0.5.
2. Leave-many-out (LMO) cross-validation
   The basic principle of the LMO technique or leave-some-out (LSO) technique is that a definite portion of the training set is held out and eliminated in each cycle. For each cycle, the model is constructed based on the remaining molecules (and using the originally selected descriptors) and then the activity of the deleted

compounds is predicted using the developed model. After all the cycles have been completed, the predicted activity values of the compounds are used for the calculation of the LMO-Q^2.

3. True Q^2

   Hawkins et al. [11] proposed the concept of 'true Q^2' parameter, calculated based on application of the variable selection strategy at each validation cycle. The parameter may be a better tool for assessing model predictivity, chiefly in the case of small data sets, compared to the traditional approach of the splitting of the data set into training and test sets.

4. The r_m^2 metric for internal validation

   An acceptable value of Q^2 does not inevitably indicate that the predicted activity data lie in close propinquity to the observed ones although there may exist a good overall correlation between the values. Thus, to obviate this problem and to better indicate the model predictability, the r_m^2 metrics introduced by Roy et al. [12] may be computed by the following equations:

$$\overline{r_m^2} = \frac{\left(r_m^2 + r_m'^2\right)}{2} \tag{2.10}$$

$$\Delta r_m^2 = \left| r_m^2 - r_m'^2 \right| \tag{2.11}$$

Here, $r_m^2 = r^2 \times \left(1 - \sqrt{\left(r^2 - r_0^2\right)}\right)$ and $r_m'^2 = r^2 \times \left(1 - \sqrt{\left(r^2 - r_0'^2\right)}\right)$. The parameters r^2 and r_0^2 are the squared correlation coefficients between the observed and (leave-one-out) predicted values of the compounds with and without intercept, respectively. The parameter $r_0'^2$ bears the same meaning but uses the reversed axes.

The $\overline{r_m^2}$ is the average value of r_m^2 and $r_m'^2$, and Δr_m^2 is the absolute difference between r_m^2 and $r_m'^2$. In case of internal validation of the training set, the $\overline{r_m^2}_{(LOO)}$ and $\Delta r_m^2_{(LOO)}$ parameters can be employed and it has been shown that the value of $\Delta r_m^2_{(LOO)}$ should preferably be lower than 0.2 provided that the value of $\overline{r_m^2}_{(LOO)}$ is more than 0.5. Roy et al. [13] proposed that the calculation of the r_m^2 metrics should be based on the scaled values of the observed and the predicted response data. The scaling may be done based on the following equation.

$$\text{Scaled } Y_i = \frac{Y_i - Y_{\min(obs)}}{Y_{\max(obs)} - Y_{\min(obs)}} \tag{2.12}$$

Here, Y_i refers to the observed/predicted response for the ith (1, 2, 3, ..., n) compound in the training/test set. Besides these, $Y_{\max(obs)}$ and $Y_{\min(obs)}$ indicate the maximum and minimum values, respectively, for the observed response in the training set compounds.

To make the calculation of r_m^2 metrics easier, a web application known as 'r_m^2 calculator' (http://aptsoftware.co.in/rmsquare) has been also developed.

5. True r_m^2 (LOO)

In case of LOO-CV, r_m^2 is calculated based on the LOO-predicted activity values of the training set and the parameter is referred to as r_m^2 (LOO), while the true r_m^2 (LOO) value is obtained from the model developed from the undivided data set after the application of variable selection strategy at each cycle of validation [14]. The 'true r_m^2 (LOO)' metric may reflect characteristics of external validation without loss of chemical information.

6. Metrics for chance correlation: Y-randomization

Y-randomization is performed in order to ensure the robustness of the developed QSAR model. In the Y-randomization test, validation is performed by permuting the response values (Y) with respect to the X matrix which has been kept unaltered. This method is generally performed in two different ways: (a) process randomization and (b) model randomization performed at varying confidence levels. The deviation in the values of the squared mean correlation coefficient of the randomized model (R_r^2) from the squared correlation coefficient of the non-random model (R^2) is reflected in the value of $^cR_p^2$ parameter computed from the following equation [15]:

$$^cR_p^2 = R \times \sqrt{R^2 - R_r^2} \qquad (2.13)$$

The threshold value of $^cR_p^2$ is 0.5. For a QSAR model having the corresponding value above the stated limit, it might be considered that the model is not obtained by chance only.

2.3.3.2 Metrics for External Validation

1. Predictive R^2 $\left(R_{pred}^2 \text{ or } Q_{(F1)}^2 \right)$

The R_{pred}^2 reflects the degree of correlation between the observed and predicted activity data of the test set.

$$R_{pred}^2 = 1 - \frac{\sum \left(Y_{obs(test)} - Y_{pred(test)} \right)^2}{\sum \left(Y_{obs(test)} - \overline{Y}_{training} \right)^2} \qquad (2.14)$$

Here, $Y_{obs(test)}$ and $Y_{pred(test)}$ are the observed and predicted activity data for the test set compounds, while $\overline{Y}_{training}$ indicates the mean observed activity of the training set molecules. Thus, models with values of R_{pred}^2 above the stipulated value of 0.5 are considered to be well predictive.

2. Golbraikh and Tropsha's criteria

Golbraikh and Tropsha [16] proposed a set of parameters for determining the external predictability of QSAR model. According to Golbraikh and Tropsha, models are considered satisfactory, if all of the following conditions are satisfied:

(a) $Q^2_{\text{training}} > 0.5$.

(b) $R^2_{\text{test}} > 0.6$.

(c) $\dfrac{r^2 - r_0^2}{r^2} < 0.1$ and $0.85 \leq k \leq 1.15$ or

$\dfrac{r^2 - r_0'^2}{r^2} < 0.1$ and $0.85 \leq k' \leq 1.15$.

(d) $\left| r_0^2 - r_0'^2 \right| < 0.3$.

The meaning of the r^2 and r_0^2 terms is already discussed in the 'r_m^2 metric for internal validation' section.

3. The $r^2_{m(\text{test})}$ metric for external validation

In order to verify the propinquity between the observed and predicted data, the parameter $r^2_{m(\text{test})}$, similar to $r^2_{m(\text{LOO})}$ used in internal validation, has been developed by Roy et al. [12]. The value of $r^2_{m(\text{test})}$ is calculated using the squared correlation coefficients between the observed and predicted activity of the test set compounds. For the acceptable prediction, the value of Δr^2_m (test) should preferably be lower than 0.2 provided that the value of $\overline{r^2_{m(\text{test})}}$ is more than 0.5.

More interestingly, Roy and coworkers established that this tool can be extended to the entire data set employing the LOO-predicted activity for the training set and predicted activity for the test set compounds. These parameters have been referred to as $\overline{r^2_{m(\text{overall})}}$ and Δr^2_m (overall) which reflect the predictive ability of the model for the entire data set.

4. RMSEP

External predictive ability of a QSAR model may further be determined by root mean square error in prediction (rmsep) given by Eq. (2.15).

$$\text{RMSEP} = \sqrt{\dfrac{\sum \left(y_{\text{obs(test)}} - y_{\text{pred(test)}} \right)^2}{n_{\text{ext}}}} \tag{2.15}$$

Here, n_{ext} refers to the number of test set compounds.

5. $Q^2_{(F2)}$

$Q^2_{(F2)}$ is based on prediction of test set compounds ($Q^2_{(F2)}$) proposed by Schüürmann et al. [17] as given by Eq. (2.16).

$$Q^2_{(F2)} = 1 - \dfrac{\sum \left(Y_{\text{obs(test)}} - Y_{\text{pred(test)}} \right)^2}{\sum \left(Y_{\text{obs(test)}} - \overline{Y}_{\text{test}} \right)^2} \tag{2.16}$$

Here, $\bar{Y}_{\text{test}}$ refers to the mean observed data of the test set compounds. A threshold value 0.5 is defined for this parameter.

6. $Q^2_{(F3)}$

The $Q^2_{(F3)}$ metric with a threshold value of 0.5, for validation of a QSAR model has been proposed by Consonni et al. [18]. This parameter is defined as follows:

$$Q^2_{(F3)} = 1 - \frac{\left[\sum \left(Y_{\text{obs(test)}} - Y_{\text{pred(test)}}\right)^2\right] / n_{\text{ext}}}{\left[\sum \left(Y_{\text{obs(train)}} - \bar{Y}_{\text{train}}\right)^2\right] / n_{\text{tr}}} \qquad (2.17)$$

where n_{tr} refers to the number of compounds in the training set. However, although the value of $Q^2_{(F3)}$ measures the model predictability, it is sensitive to training set data selection and tends to penalize models fitted to a very homogeneous data set even if predictions are close to the truth.

7. Concordance correlation coefficient (CCC)

The CCC parameter can be calculated in order to check the model reliability by the following equation [19]:

$$\bar{p}_c = \frac{2\sum_{i=1}^{n} \left(x_{\text{obs(test)}} - \overline{x_{\text{obs(test)}}}\right)\left(y_{\text{pred(test)}} - \bar{y}_{\text{pred(test)}}\right)}{\sum_{i=1}^{n} \left(x_{\text{obs(test)}} - \overline{x_{\text{obs(test)}}}\right)^2 + \sum_{i=1}^{n} \left(y_{\text{pred(test)}} - \bar{y}_{\text{pred(test)}}\right)^2 + n\left(\overline{x_{\text{obs(test)}}} - \bar{y}_{\text{pred(test)}}\right)}$$

$$(2.18)$$

In the above equation, $x_{\text{obs(test)}}$ and $y_{\text{pred(test)}}$ correspond to the observed and predicted values of the test compounds, n is the number of chemicals, and $\overline{x_{\text{obs(test)}}}$ and $\overline{y_{\text{pred(test)}}}$ correspond to the averages of the observed and predicted values, respectively, for the test compounds. The ideal value of CCC should be equal to 1.

The $r^2_{m(\text{rank})}$ metric

In order to assess the closeness between the order of the predicted activity and that of the observed activity, the $r^2_{m(\text{rank})}$ parameter was developed. The $r^2_{m(\text{rank})}$ metric is computed based on the correlation of the ranks generated for the observed and the predicted response data. An ideal ranking where the observed and the predicted response data perfectly match with each other yields zero difference between the two values for each molecule, and the $r^2_{m(\text{rank})}$ metric attains a value of unity.

$$r^2_{m(\text{rank})} = r^2_{(\text{rank})} \times \left(1 - \sqrt{r^2_{(\text{rank})} - r^2_{0(\text{rank})}}\right) \qquad (2.19)$$

2.3.4 Validation Metrics Employed in Classification-Based QSAR

Validation metrics can assess the performance of the classification-based models in terms of accurate qualitative prediction of the dependent variable. Commonly applied metrics for classification-based QSAR models are illustrated below [20]:

2.3.4.1 Parameters for Goodness-of-Fit and Quality Determination

1. Wilks lambda (λ) statistics
   The Wilks lambda is a metric for the testing of significance of a discriminant model function and determined as the ratio of within group sum of squares and total sum of squares, i.e., within-category to total dispersion.

$$\text{Wilks } \lambda = \frac{\text{Within group sum of squares}}{\text{Total sum of squares}} \tag{2.20}$$

   The Wilks lambda value spans from 0 to 1, where 0 corresponds to good level of discrimination and 1 refers to no discrimination.
2. Canonical index (R_c)
   The quantification of the strength of the relationship between the dependent and independent variables is articulated as a canonical correlation coefficient.

$$R_c = \sqrt{\frac{\lambda_i}{1 + \lambda_i}} \tag{2.21}$$

   Here, λ_i is referred as eigen value of the matrix.
3. Chi-square (χ^2)
   The quality of classification-based model is also judged using the chi-square (χ^2) statistic.

$$\chi^2 = \sum_{i=1}^{t} \frac{(f_i - F_i)^2}{F_i} \tag{2.22}$$

   where f_i is observed response, F_i is predicted response, and t is the number of observations.
4. Squared Mahalanobis distance
   The square of Mahalanobis distance is calculated for the determination of probability of a compound to be classified in a definite group in the discriminant

space for LDA. In a multivariate normal distribution with covariance matrix Σ, the Mahalanobis distance between any two data points x_i and x_j can be defined as follows:

$$d_{\text{mahalanobis}}\left(x_i, x_j\right) = \sqrt{\left(x_i - x_j\right)^T \sum{}^{-1}\left(x_i - x_j\right)} \qquad (2.23)$$

where x_i and x_j are two random data points, T is transpose of a matrix, and Σ^{-1} is inverse of the covariance matrix.

2.3.4.2 Metrics for Model Performance Parameters

1. Sensitivity, Specificity, and Accuracy
   The compounds classified employing the classification-based QSAR model can be divided into four categories based on a comparison between the predicted and observed response:

   (a) True positives (TP): the active compounds which have been correctly predicted as actives,
   (b) False negatives (FN): this class includes the active compounds which have been erroneously classified as inactives,
   (c) False positives (FP): this class comprises the inactive compounds wrongly classified as actives,
   (d) True negatives (TN): this class accounts for the inactive compounds which have been accurately predicted as inactives.

   Based on the two-by-two confusion matrix, the following metrics can be computed to evaluate the classifier model performance and classification capability.

$$\text{Sensitivity} = \text{Recall} = \frac{TP}{TP + FN} \qquad (2.24)$$

$$\text{Specificity} = \frac{TN}{TN + FP} \qquad (2.25)$$

$$\text{Accuracy} = \frac{TP + TN}{TP + FP + TN + FN} \qquad (2.26)$$

2. *F*-measure and Precision
   The *F*-measure refers to the harmonic mean of recall and precision, where recall refers to the accuracy of real prediction and precision defines the accuracy of a predicted class.

$$F\text{-measure} = \frac{2(\text{Recall})(\text{Precision})}{\text{Recall} + \text{Precision}} \qquad (2.27)$$

$$\text{Precision} = \frac{\text{TP}}{\text{TP} + \text{FP}} = \text{fp rate} \qquad (2.28)$$

3. *G*-means

Combining sensitivity and specificity into a single parameter via the geometric mean (*G*-means) allows for a straightforward way to assess the model's ability to perfectly classify active and inactive samples using the formula:

$$G\text{-means} = \sqrt{\text{Sensitivity} \times \text{Specificity}} \qquad (2.29)$$

4. Cohen's κ

Cohen's kappa (κ) can be employed to determine the agreement between classification (predicted) models and known classifications. It can be defined as follows:

$$\text{Cohen's } \kappa = \frac{P_r(a) - P_r(e)}{1 - P_r(e)} \qquad (2.30)$$

$$P_r(a) = \frac{(\text{TP} + \text{TN})}{(\text{TP} + \text{FP} + \text{FN} + \text{TN})} \qquad (2.31)$$

$$P_r(e) = \frac{\{(\text{TP} + \text{FP}) \times (\text{TP} + \text{FN})\} + \{(\text{TN} + \text{FP}) \times (\text{TN} + \text{FN})\}}{(\text{TP} + \text{FN} + \text{FP} + \text{TN})^2} \qquad (2.32)$$

Here, $P_r(a)$ is the relative observed agreement between the predicted classification of the model and the known classification, and $P_r(e)$ is the hypothetical probability of chance agreement. Cohen's kappa analysis returns values between −1 (no agreement) and 1 (complete agreement).

5. Matthews correlation coefficient (MCC)

The MCC is regarded as a balanced measure which can be employed even if the classes are of diverse sizes. The MCC is simply a correlation coefficient between the observed and predicted binary classifications, and it returns a value between −1 and +1. A coefficient of +1 signifies a perfect prediction, 0 an average random prediction, and −1 an inverse prediction. The MCC can be computed directly from the confusion matrix using the formula:

$$MCC = \frac{TP \times TN - FP \times FN}{\sqrt{(TP + FP)(TP + FN)(TN + FP)(TN + FN)}} \tag{2.33}$$

The meaning of TP, TN, FP, and FN are already discussed.

2.3.5 Parameters for Receiver Operating Characteristics (ROC) Analysis

1. ROC curve
   The ROC curve is a visual illustration of the success and error observed in a classification model. The curve is plotted taking true positive rate (tp) on the y-axis and false-positive rate (fp) on the x-axis, and the characteristics of the curve provides easier recognition of the precision of prediction [21].

$$\text{tp rate} \approx \frac{\text{Positives (active molecules) correctly classified}}{\text{Total postives}} = \text{Sensitivity} \tag{2.34}$$

$$\text{fp rate} = \frac{\text{Negatives (inactive compounds) incorrectly classified}}{\text{Total negatives}}$$
$$= 1 - \text{specificity} \tag{2.35}$$

The ROC curve signifies the number of objects the classifier identifies correctly as well as the number wrongly identified by the classifier.

2. ROCED and ROCFIT
   Two metrics based on distances in a ROC curve for the selection of classification models with an correct balance in both training and test sets, namely the ROC graph Euclidean distance (ROCED) and the ROC graph Euclidean distance corrected with fitness function (FIT(λ)) or Wilks λ (ROCFIT), are also used [22]. The Euclidean distance between the perfect and a real classifier (d_i) expressed as a function of their respective values of sensitivity and specificity is

$$d_1 = \sqrt{\left(Se_p - Se_r\right)^2 + \left(Sp_p - Sp_r\right)^2} \tag{2.36}$$

where Se_p and Se_r are the respective sensitivity values of the perfect and the real classifier, while Sp_p and Sp_r represent the specificity values of the perfect and real classifier, respectively. Since the sensitivity and specificity for a perfect classifier takes values of 1, the Euclidean distance can be expressed as

$$d_1 = \sqrt{(1 - \text{Se}_r)^2 + (1 - \text{Sp}_r)^2} \tag{2.37}$$

$$\text{ROCED} = (|d_1 - d_2| + 1) \times (d_1 + d_2) \times (d_2 + 1) \tag{2.38}$$

where d_1 and d_2 are representation of the distances in a ROC graph for the training and test sets, respectively. ROCED takes values between 0 (perfect classifier) and 4.5 (random classifier).

A new parameter ROCFIT has also been introduced. ROCFIT is defined as follows:

$$\text{ROCFIT} = \frac{\text{ROCED}}{\text{Wilks}(\lambda)} \tag{2.39}$$

2.3.5.1 Metrics for Pharmacological Distribution Diagram (PDD)

The PDD is a frequency distribution plot of a dependent variable where expectancy values of the variable are plotted in the y-axis against numeric intervals of the variable in the x-axis [23]. This graph visually signifies the overlapping regions of the categories, e.g., positives and negatives. For a classification case comprising two classes such as actives and inactives (or positives and negatives), two terms named 'active expectancy' and 'inactive expectancy' may be defined as below where the denominator is added with a numerical value of 100 to avoid division by zero:

$$\text{Activity expectancy} = E_a = \frac{\text{Percentage of actives}}{(\text{Percentage of inactives} + 100)} \tag{2.40}$$

$$\text{Inactivity expectancy} = E_i = \frac{\text{Percentage of inactives}}{(\text{Percentage of actives} + 100)} \tag{2.41}$$

where 'a' and 'i' are the number of occurrences of active and inactive compounds at a specific range.

2.4 Conclusion

The QSAR/QSPR modeling technique involves the use of a significant number of statistical tools and hence requires a good knowledge of chemometrics. The developed QSAR model can furnish linear as well as nonlinear relationship between the response and chemical attributes through regression-based as well as classification-based analyses. Since, quantitative mathematical relationships are established, validation of the models using a suitable statistical algorithm becomes essential to confirm the stability and predictivity of the models. The judgment for the choice of method depends upon a multitude of factors including the response to be modeled, the nature of the training set data, the type of descriptors used and also its numbers, and even the objective of the analysis.

References

1. Snedecor GW, Cochran WG (1967) Statistical methods. Oxford and IBH, New Delhi
2. Wold S, Sjöström M, Eriksson L (2001) PLS-regression: a basic tool of chemometrics. Chemom Intell Lab Syst 58:109–130
3. Agresti A (1996) An introduction to categorical data analysis. Wiley, Hoboken
4. Everitt BS, Landau S, Leese M (2001) Cluster analysis, 4th edn. Arnold, London
5. Topliss JG, Costello RJ (1972) Chance correlation in structure-activity studies using multiple regression analysis. J Med Chem 15:1066–1068
6. Jaworska JS, Comber M, Auer C, Van Leeuwen CJ (2003) Summary of a workshop on regulatory acceptance of (Q)SARs for human health and environmental endpoints. Environ Health Perspect 111:1358–1360
7. Wold S (1978) Cross-validation estimation of the number of components in factor and principal components models. Technometrics 20:397–405
8. Roy K (2007) On some aspects of validation of predictive QSAR models. Expert Opin Drug Discov 2:1567–1577
9. Gramatica P (2007) Principles of QSAR models validation: internal and external. QSAR Comb Sci 26:694–701
10. Roy K, Mitra I (2011) On various metrics used for validation of predictive QSAR models with applications in virtual screening and focused library design. Comb Chem High Throughput Screen 14:450–474
11. Hawkins DM, Basak SC, Mills D (2003) Assessing model fit, by cross-validation. J Chem Inf Comput Sci 43:579–586
12. Roy K, Mitra I, Kar S, Ojha PK, Das RN, Kabir H (2012) Comparative studies on some metrics for external validation of QSPR models. J Chem Inf Model 52:396–408
13. Roy K, Chakraborty P, Mitra I, Ojha PK, Kar S, Das RN (2013) Some case studies on application of "r_m^2" metrics for judging quality of QSAR predictions: emphasis on scaling of response data. J Comput Chem 34:1071–1082
14. Mitra I, Roy PP, Kar S, Ojha P, Roy K (2010) On further application of rm2 as a metric for validation of QSAR models. J Chemometrics 24:22–33
15. Mitra I, Saha A, Roy K (2010) Exploring quantitative structure-activity relationship (QSAR) studies of antioxidant phenolic compounds obtained from traditional Chinese medicinal plants. Mol Simult 36:1067–1079
16. Golbraikh A, Tropsha A (2002) Beware of q2! J Mol Graph Model 20:269–276
17. Schuurmann G, Ebert RU, Chen J, Wang B, Kuhne R (2008) External validation and prediction employing the predictive squared correlation coefficient-Test-set activity mean vs training set activity mean. J Chem Inf Model 48:2140–2145
18. Consonni V, Ballabio D, Todeschini R (2010) Evaluation of model predictive ability by external validation techniques. J Chemometrics 24:194–201
19. Chirico N, Gramatica P (2011) Real External predictivity of QSAR models: How to evaluate it? Comparison of different validation criteria and proposal of using the concordance correlation coefficient. J Chem Inf Model 51:2320–2335

20. Roy K, Kar S (2014) How to judge predictive quality of classification and regression based QSAR models? In: Haq Z, Madura JD (eds) Frontiers in computational chemistry. Bentham Science Publishers, Sharjah

21. Fawcett T (2006) An introduction to ROC analysis. Pattern Recognit Lett 27:861–874

22. Perez-Garrido A, Helguera AM, Borges F, Cordeiro MNDS, Rivero V, Escudero AG (2011) Two new parameters based on distances in a receiver operating characteristic chart for the selection of classification models. J Chem Inf Model 51:2746–2759

23. Galvez J, Garcia-Domenech R, de Gregorio Alapont C, De Julian-Ortiz V, Popa L (1996) Pharmacological distribution diagrams: a tool for de novo drug design. J Mol Graph 14:272–276

Chapter 3
QSAR/QSPR Methods

Abstract QSAR/QSPR analysis started with different classical approaches constituting the core concept of predictive modeling analysis in the context of structure–activity relationships. Such classical techniques have been based on various postulates and hypotheses. With the passage of time, various dimensional features have taken an important role in diagnosis of chemical information and thereby in the development of successful QSAR/QSPR models. Development of computer technology has provided an essential support for easy and accurate implementation of complex molecular modeling calculations and data generation. The present chapter provides an account of the classical QSAR/QSPR approaches along with glimpses of two- and three-dimensional QSAR/QSPR techniques. The impact of the usage of computer and computational chemistry techniques in the paradigm of QSAR/QSPR has also been discussed.

Keywords 3D-QSAR · CoMFA · CoMSIA · Free–Wilson model · Fujita–Ban modification · LFER model · MSA · Topology · Graph theory · Simulation · Molecular mechanics · Density function theory · Quantum mechanics

3.1 Introduction

The QSAR/QSPR research in its present form stemmed originally from the classical approaches of Hansch analysis and de novo mathematical models of Free and Wilson. There are several prerequisites which should be met before a QSAR/QSPR analysis on a particular data set of chemical compounds can be performed.

1. The biological activity data should be of equiresponse type. If it is not so, the data should be adjusted to obtain this effect.
2. The biological activity of all the compounds under consideration should have been measured under the same conditions.
3. The congeners used for model development should be closely similar to ensure same mechanism of action for all compounds.

© The Author(s) 2015
K. Roy et al., *A Primer on QSAR/QSPR Modeling*,
SpringerBriefs in Molecular Science, DOI 10.1007/978-3-319-17281-1_3

4. The contributions of the substituent groups to the selected response should be intrinsically additive.
5. It is also desirable to use a low number of descriptors and a large number of data points allowing maximum degrees of freedom and higher statistical significance.
6. Some compounds may be required to be omitted from the data set due to outlier behavior (showing a large difference between the observed and calculated values).
7. Precaution should be taken in selection of the data to avoid ill-conditioned matrices.

Different dimensions have a good impact toward the development of predictive QSAR models. Dimension, in mathematical language, refers to the number of coordinates employed for identifying an object in it. Dimension in QSAR study provides chemical information of a molecule and aids in the development of quantitative descriptors. Following classical methods of QSAR analysis, implementation of various dimensional perspectives has explored means for structural diagnosis and the computed attributes thereof. Graph theoretical approach based on two-dimensional basis is one such unique method of molecular representation. The descriptors derived from this approach are known as topological parameters. The implementation of chemical graphs for deriving topological descriptors started in the middle of twentieth century. In the 1980s, the concept of three-dimensional analysis in relation to QSAR modeling emerged and a more realistic picture of the molecular environment was obtained. QSAR methods using two-dimensional features are principally ligand-based, while the three-dimensional attributes allow ligand as well as structure-based analysis. Even, various higher dimensional methods (i.e., more than three) are nowadays exercised by the researchers. It will be noteworthy to mention that exploration of computer technology has provided an essential platform to QSAR analysis by allowing complex molecular simulation operations. Accurate computation of simple-to-complex descriptors can be easily carried out using suitable software algorithm in computers. Furthermore, the visualization graphics in computer also helps in better molecular understanding.

3.2 De Novo Models

De novo QSAR models are the mathematical models which do not require computation of any descriptors encoding chemical information on molecular structure. Indicator parameters (having a binary value 0 or 1) representing presence or absence of a group at a particular position are used for development of the models.

3.2.1 Free–Wilson Model

In 1964, Free and Wilson [1] developed an additive mathematical model based on the measurement of contributions of different substituents at specified positions of a congeneric series of compounds to the biological activity. The original model did not use logarithmic transformation of the biological activity as is done in the current QSAR practice. For Free–Wilson model development, the data set should be a congeneric series having substituents at specified (and at least two) positions. Additionally, a particular substituent should occur at least twice at a particular position of the data set. The basic assumption of the Free–Wilson model is that the contribution of a particular group at a specified position of the congeneric series of compounds is same in all such compounds without considering the cross-interaction terms. Mathematically, Free–Wilson model can be represented in the following expression [2, 3]:

$$\text{BA} = \sum G_i X_i + \mu \tag{3.1}$$

In the above equation, BA is biological activity, G_i is the contribution of a particular group i, while X_i indicates presence or absence (value 1 or 0) of a particular group. The constant μ is the contribution of the parent moiety. The net contribution of all the substituents occurring at a particular position is considered zero (this is known as symmetry restriction). This constraint helps to achieve unique solutions for the substituent constants.

3.2.2 Fujita–Ban Model

This is a modification [4] of the original Free–Wilson model. It differs from the Free–Wilson model in three aspects. In the Fujita–Ban method, the activity contribution of a substituent relative to that of 'H' at each position is considered unlike the Free–Wilson model, where 'H' is considered as a substituent to the parent moiety. This obviates the requirement of symmetry equations in the Fujita–Ban model, thus simplifying calculations. Moreover, in the Fujita–Ban model, the constant term signifies the response value of the unsubstituted compound, while in the Free–Wilson model, it is the contribution of the parent moiety. Finally, the Free–Wilson model does not use log-transformed response value, while log of activity is considered as the response in the Fujita–Ban model. Mathematically, the Fujita–Ban model can be expressed as the following:

$$\log A = \sum G_i X_i + \log A_0 \tag{3.2}$$

In the above equation, $\log A$ is the log-transformed activity of the substituted compound, while $\log A_0$ is the log-transformed activity of the unsubstituted compound. G_i is the contribution of the ith substituent to the activity relative to H and X_i is a binary variable having a value 1 (presence of ith substituent) or zero (absence of ith substituent).

The advantage of the Free–Wilson and Fujita–Ban models is that they do not require computation of any descriptors. Information on mere presence or absence of groups at particular position can lead to development of the models which may give the first-hand information about the trend of structure–activity relationship. However, the problem of these models is that they cannot be used for prediction of activity of the compounds containing substituents which are not present in the modeling set.

3.3 Property-Based QSAR

3.3.1 LFER Approach of Hansch

Physicochemical properties of chemical compounds have been widely used as descriptors in QSAR/QSPR studies. There are three main categories of physico-chemical properties (for either whole molecules or substituents) used for modeling: hydrophobic, electronic, and steric. The property-based QSAR approach was originally developed and promoted to the medicinal chemists by Prof. C Hansch through his linear free energy-related (LFER) approach.

The LFER approach of Hansch using physicochemical descriptors and sub-stituent constants has its origin in the work of Hammett [5] in physical organic chemistry.

Hammett defined an electronic substituent constant σ for the hydrolysis rates of benzoic acid derivatives in the following expression:

$$\log\left(K_X/K_H\right) = \rho\sigma \tag{3.3}$$

In the above equation, K_X and K_H are the equilibrium constants (or rate constants) for the reactions of substituted and unsubstituted benzoic acids, respectively, ρ is a constant dependent on type and conditions of the reaction as well as the nature of compounds, σ is an electronic substituent constant depending on its nature and position of the substituent. Equation (3.3) may be rewritten as

$$\log K_X = \rho\sigma + \log K_H \tag{3.4}$$

Note that Hammett σ is applicable for *meta-* and *para-*aromatic substituents. In analogy to the Hammett σ equation, Hansch and Fujita [6] introduced another substituent constant π in the following manner:

$$\pi_X = \log\left(\frac{P_X}{P_H}\right) \tag{3.5}$$

In the above equation, π_X is the hydrophobic substituent constant of substituent X, while P_X and P_H are (n-octanol–water) partition coefficients of substituted and unsubstituted compounds.

Hansch observed a parabolic dependence of the biological activity (Fig. 3.1) on the hydrophobicity or hydrophobicity constant.

$$\log \frac{1}{C} = a\pi - b\pi^2 + c \tag{3.6}$$

or

$$\log \frac{1}{C} = a\log P - b(\log P)^2 + c \tag{3.7}$$

Equation (3.7) uses the hydrophobicity term $\log P$ for the whole molecules.

On using both electronic and hydrophobic substituent constant terms, a generalized expression of Hansch equation can be shown as follows:

$$\log \frac{1}{C} = k_1\pi - k_2\pi^2 + k_3\sigma + k_4 \tag{3.8}$$

Fig. 3.1 A parabolic relationship of the biological activity with $\log P$ (or π)

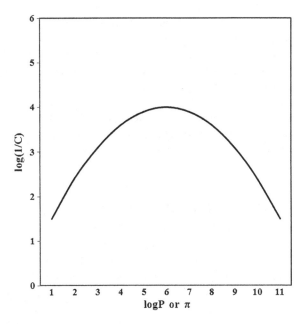

Additional terms may be added to the above expression. For example, when a steric E_s is important, Eq. (3.8) may take the following form:

$$\log {}^1/_C = k_1 \pi - k_2 \pi^2 + k_3 \sigma + k_4 E_s + k_5 \qquad (3.9)$$

All descriptors appearing in the final model should have statistically significant regression coefficients; otherwise, such terms should be omitted. In selecting the physicochemical parameters to be used in the QSAR models, one should check the possibility of intercorrelation among various pairs of substituent constants.

The Hansch model is a very general approach, as any kind of drug–receptor interactions is caused by factors which can be broadly categorized into any one or more of hydrophobic, electronic, and steric factors. The Hansch model is applicable for closely related congeners and a given biological activity. It is based on the following postulates:

(i)   The drug molecules reach the receptor site via a 'random walk' process.
(ii) The drug molecules bind with the receptor forming a complex.
(iii) The drug–receptor complex undergoes a chemical reaction or conformational change for the desired activity.
(iv) The drugs in a congeneric series should act through the same mechanism of action.

The descriptors (physicochemical properties for whole molecules or substituents) in the Hansch model have values in a continuous scale; thus, this approach may be used for prediction of the response for compounds having such substituents not present in the modeling set.

The Hansch approach has been very successful in QSAR studies of drugs and other biologically active chemicals. There are many successful applications of this approach reported in the literature [7]. By approximating the physicochemical properties with measured or theoretical values, one may be able to use the method as a measure to determine the relative importance and role of each factor in the biological mechanism. However, this approach is applicable only to closely related congeners sharing a common mechanism of action.

3.3.2 The Mixed Approach

The Hansch approach and the Fujita–Ban model can be combined to a mixed approach. If for one definite region of the molecule, a Hansch correlation can be obtained for the substituents, while substituents in another position of the molecule must be treated by Free–Wilson analysis (using indicator variables for the presence or absence of substituents at particular positions), the Fujita–Ban model and the Hansch approach can be combined to a mixed approach as given in the following expression [7]:

$$\log {^1/_C} = k_1 \pi + k_2 \sigma + \sum G_i X_i + c \qquad (3.10)$$

In Eq. (3.10), $k_1 \pi + k_2 \sigma$ is the Hansch part for the substituents Y_j and $\sum G_i X_i$ is the (modified) Free–Wilson part for the substituents X_i (with G_i being corresponding group contributions), while c is the theoretically predicted activity value of the unsubstituted parent compound ($X = Y = H$) or of an arbitrarily chosen reference compound.

3.4 Graph Theoretical Approach

3.4.1 Introduction to Graph Theory

The formalism of QSAR modeling studies began with the use of physicochemical properties as descriptors which was followed by the application of graph theoretical concept. The concept of graph originates from mathematics and usually confers to the collection of a set of objects in a plane and their binary relationship. The inclusion of this mathematical concept into chemistry enables depiction of chemical objects in plane which are atoms, bonds, groups, etc. Hence, the mathematical graphs are transcribed to 'chemical graphs' when information of molecular structures is used. The two basic elements of chemical graphs are 'vertex' and 'edge' which depict a connected molecular structure. Atoms are represented by vertex, while edges correspond to covalent chemical bonds. This process enables a special pattern of representation of molecular structure known as hydrogen-suppressed or hydrogen-depleted molecular graphs meaning that a molecular structure is to be represented using vertices and edges without the portrayal of explicit hydrogen atoms [8]. In Fig. 3.2, it is shown that the hydrogen-suppressed molecular graph of isopentane comprises five vertices (atoms) and four edges (bonds). We can see that graph theoretical formalism provides a two-dimensional molecular representation and the information derived thereof will primarily be of the same nature. Furthermore, because of the two-dimensional nature, the graph theoretical molecular representation does not require computation of any specific bond length or angle. It may be noted that simple dots or points are used for vertex representation of carbon atoms, while heteroatoms are represented by their symbols (with added H-atom if any like –OH, –NH$_2$, etc.).

3.4.2 Matrix and Chemical Graphs

The use of chemical graph theory became a useful concept in the paradigm of QSAR analysis when the graphical depiction of molecular structure was incorporated into a mathematical matrix that led into the derivation of suitable molecular

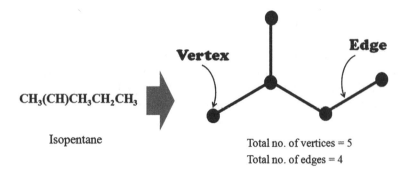

CH₃(CH)CH₃CH₂CH₃

Isopentane

Total no. of vertices = 5
Total no. of edges = 4

Fig. 3.2 Hydrogen-suppressed chemical graph of isopentane molecule showing its vertex and edge

descriptors. Matrices can be identified as the arrays of numbers or some other mathematical objects which can be used as an abstract platform for the algorithmic encoding of a desired problem. In the context of graph theoretical chemistry, matrices are developed using the connectivity information of chemical structures and are treated with suitable algebraic operators to derive two-dimensional descriptors known as topological indices or topological descriptors. Therefore, a matrix enables the codification of chemical graphs in the form of numbers which might be subjected to operators for the derivation of topological descriptors [9, 10].

The formation of a graph theoretical matrix is based on the bonded connections between atoms. By the use of a suitable formalism such as adjacency and distance, the connectivity in a molecular graph can be in the form of numbers which are used as the elements of a matrix. Figure 3.3 shows the formation of a distance matrix (vertex-based) for the molecule isobutane and the corresponding steps taken are depicted as follows:

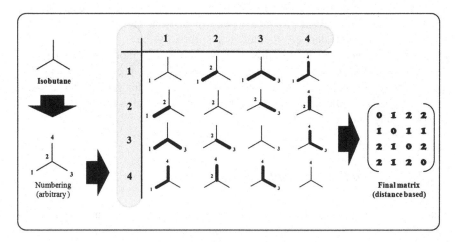

Fig. 3.3 An example of topological distance-based matrix formation for the molecule isobutane

(i) Drawing of the chemical structure of the molecule using vertices and edges without showing the hydrogen atoms. For the isobutane molecule, the number of vertices (here carbon atom) is four, while the number of edges is three (sigma bonds)
(ii) An arbitrary numbering of the vertex elements
(iii) Determination of the desired distance or adjacency information
(iv) Development of the matrix using the counted values

Two types of matrices, namely distance-based and adjacency-based, are commonly encountered in defining chemical graph theory-related problems although another miscellaneous group can be identified. Each type of matrices further comprises several subgroups which are treated with mathematical operators to give single quantitative information. In the section below, the formal definition along with the examples of some representative graph theoretical matrices is presented [8–11].

(a) Vertex-adjacency matrix: The vertex-adjacency matrix of a connected molecular graph G can be defined as follows:

$$[A_v(G)]_{ij} = 1 \quad \text{when } i \neq j \text{ and } e_{ij} \in E(G) \text{ i.e., vertices } i \text{ and } j \text{ are adjacent}$$
$$= 0 \quad \text{when } i = 0 \text{ and } e_{ij} \notin E(G)$$

$$(3.11)$$

where e_{ij} is the edge defined by the vertices i and j, and $E(G)$ is the set of edges present in the connected molecular graph G. Figure 3.4 shows the elements of vertex-adjacency matrix for the compound n-propane.

(b) Edge-adjacency matrix: The edge-adjacency matrix of a connected graph G may be represented as follows:

$$[A_e(G)]_{ij} = 1 \quad \text{when } e_{ij} \in E(G) \text{ i.e., edges } i \text{ and } j \text{ are adjacent}$$
$$= 0 \quad \text{when } e_{ij} \notin E(G)$$

$$(3.12)$$

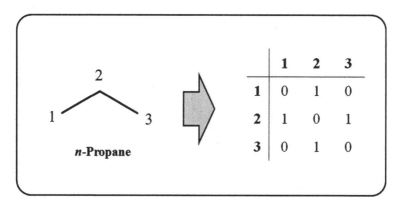

Fig. 3.4 Vertex-adjacency matrix elements for the molecule n-propane

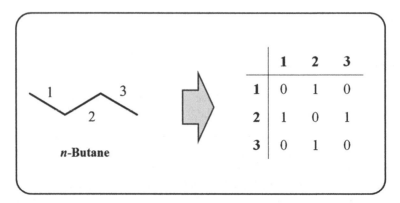

Fig. 3.5 Edge-adjacency matrix elements for the molecule n-butane

where edge e_{ij} is constituted by vertices i and j in a molecule comprising of E (G) set of graph edges. Figure 3.5 shows the vertex-adjacency matrix elements for the sample compound n-butane.

Vertex-adjacency matrix enables the differentiation of specific type of graph unlike edge-adjacency matrix that could not separate between non-isomorphic graphs.

(c) Vertex-distance matrix: For a connected molecular graph G, the vertex-distance matrix can be defined as follows:

$$
\begin{aligned}
[D(G)]_{ij} &= (d_{ij})_{\min} \qquad \text{if } i \neq j \\
&= 0 \qquad\qquad\ \text{if } i = j
\end{aligned}
\tag{3.13}
$$

where $(d_{ij})_{\min}$ denotes the minimum topological distance between vertices i and j. The vertex-distance matrix elements for the compound isobutane are shown in Fig. 3.6.

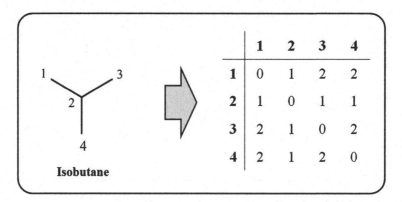

Fig. 3.6 Vertex-distance matrix elements for the molecule isobutane

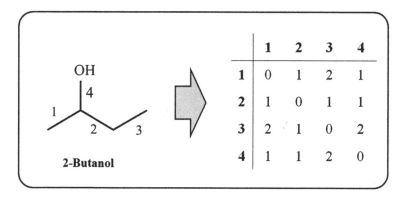

Fig. 3.7 Edge distance matrix elements for the molecule 2-butanol

(d) Edge-distance matrix: The edge-distance matrix for a connected molecular graph G may be defined as follows:

$$[D(G)]_{ij} = (d_{ij})_{min} \quad \text{if } i \neq j$$
$$= 0 \quad \text{if } i = j \tag{3.14}$$

where the minimum topological distance between edges i and j is denoted by $(d_{ij})_{min}$. Figure 3.7 denotes the elements for the edge-adjacency matrix of the sample compound 2-butanol.

Hence, we can see that similar formalism can be used for deriving matrices based on the count of vertex as well as edge of chemical graphs. In the next few examples, the derived matrices are based on the features of the graph vertices.

(e) Distance-complement matrix: It may be defined as depicted below.

$$[D_c(G)]_{ij} = V - [D(G)]_{ij} \quad \text{if } i \neq j$$
$$= 0 \quad \text{if } i = j \tag{3.15}$$

where V represents the number of vertices in a connected molecular graph G. Figure 3.8 depicts an element of the distance-complement matrix for the compound n-pentane.

(f) Reciprocal distance matrix: This matrix is also known as 'Harary matrix (or vertex-Harary matrix)' and can be defined as follows:

$$[RD(G)]_{ij} = 1/[D(G)]_{ij} \quad \text{if } i \neq j$$
$$= 0 \quad \text{if } i = j \tag{3.16}$$

where the topological graph distance between vertices i and j is denoted by $[D(G)]_{ij}$. Elements of the reciprocal distance matrix for the sample molecule 2-methylpentane has been furnished in Fig. 3.9.

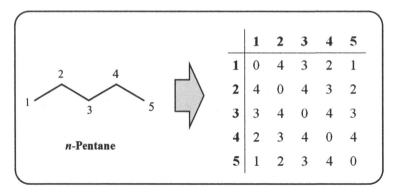

Fig. 3.8 Distance-complement matrix elements for the molecule n-pentane

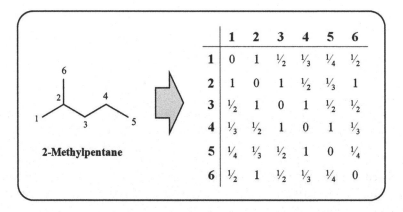

Fig. 3.9 Reciprocal distance matrix elements for the molecule 2-methylpentane

(g) Distance-path matrix: It may be defined as follows.

$$
\begin{aligned}
\left[D_p(G)\right]_{ij} &= [D(G)]_{ij}\left([D(G)]_{ij}+1\right)\big/2 \quad \text{if } i \neq j \\
&= 0 \qquad\qquad\qquad\qquad\qquad\quad \text{if } i = j
\end{aligned}
\tag{3.17}
$$

where $D(G)_{ij}$ denotes the distance matrix element of vertices i and j. Here, all possible internal paths between vertices i and j are considered in the matrix $D_p(G)_{ij}$. A sample distance-path matrix element for the molecule 2,3-dimethylbutane has been presented in Fig. 3.10.

(h) Detour matrix: This matrix is derived using path length in a connected molecular graph. The longest possible distance in a molecular graph is designated as 'detour distance' which is also known as the elongation. The matrix may be defined as follows:

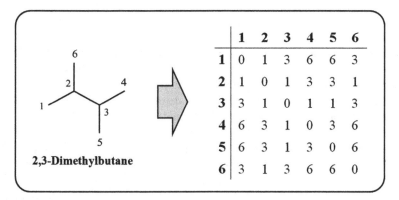

Fig. 3.10 Distance-path matrix elements for the molecule 2,3-dimethylbutane

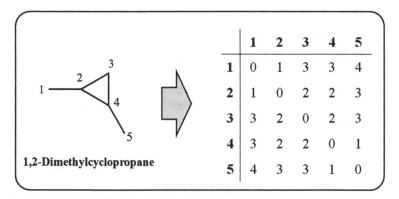

Fig. 3.11 Detour matrix elements for the molecule 1,2-dimethylcyclopropane

$$[\Delta(G)]_{ij} = \max(l(p_{ij})) \quad \text{if } i \neq j \atop = 0 \qquad\qquad\quad \text{if } i = j \tag{3.18}$$

where the path is denoted by p_{ij}, $l(p_{ij})$ corresponds to the length of the path, and $\max(l(p_{ij}))$ corresponds to the longest path length between vertices i and j. Figure 3.11 depicts a representative example of the detour matrix elements for the compound 1,2-dimethylcyclopropane.

It may be observed that the diagonal elements of an adjacency as well as distance-based matrix give zero value.

3.4.3 Topological Descriptors

Topological descriptors are the numerical quantities derived from graph theoretical matrices. The phrase 'topology' brings a new ground of molecular perception in theoretical chemistry by allowing a suitable mode of molecular encryption. It will be noteworthy to mention that topological descriptors are the first theoretically derived predictor variables used in QSAR modeling analysis. The perception of topology instigates from mathematics and bears conceptual similarity with 'rubber sheet geometry,' i.e., the surface of a topological object retains its property like rubber even after the application of forces, viz. twist, bend, and pull but not tearing of course [10–12]. Topological descriptors enable preservation of the properties bearing an identical value for the isomorphic graphs. Topological descriptors can be classified into two major groups as follows [13]:

(a) Topostructural indices: They provide emphasis on adjacency and graph the-
      oretical distance among participating atoms.
(b) Topochemical indices: Along with topology, such indices judge other chem-
      ical attributes, namely atom identity, hybridization state, number of core or
      valence electron.

Figure 3.12 shows the determination of topological distance for two representative molecules. The journey of topological descriptors started with Wiener index
and Platt index both of which were derived in 1947 and employed for QSPR
modeling on properties of paraffin hydrocarbons. An account of some commonly
used representative descriptors derived using topological formalism has been presented in the section of Descriptors in Chap. 2. Here, in Table 3.1 we have

Fig. 3.12 Method of determining topological distance for representative molecules

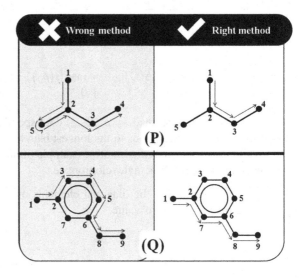

Table 3.1 Computation of values for some representative topological descriptors

Sl. No	Name of the descriptor	Mathematical formula	Sample molecule	Calculation
1	Wiener index	$W = \frac{1}{2}\sum_{i=1}^{N}\sum_{j=1}^{N}\delta_{ij}$ δ_{ij} represents the topological distance connecting vertices i and j, while N is the total number of vertices	2-Methyl-but-2-ene	Distance matrix elements: $\begin{array}{c\|ccccc\|c} & 1 & 2 & 3 & 4 & 5 & \Sigma \\ \hline 1 & 0 & 1 & 2 & 2 & 3 & 8 \\ 2 & 1 & 0 & 1 & 1 & 2 & 5 \\ 3 & 2 & 1 & 0 & 2 & 3 & 8 \\ 4 & 2 & 1 & 2 & 0 & 1 & 6 \\ 5 & 3 & 2 & 3 & 1 & 0 & 9 \\ \hline & & & & & \Sigma & 36 \end{array}$ Wiener index $= \frac{1}{2} \times 36 = 18$
2	Platt number	$F = \sum_{i=1}^{B}\sum_{j=1}^{B}[E]_{ij}$ B is the total number of edges (i.e., bonds), and E represents an edge	3-Methyl-pent-1-yne	Edge-adjacency matrix elements: $\begin{array}{c\|ccccc\|c} & 1 & 2 & 3 & 4 & 5 & \Sigma \\ \hline 1 & 0 & 1 & 0 & 0 & 0 & 1 \\ 2 & 1 & 0 & 1 & 0 & 1 & 3 \\ 3 & 0 & 1 & 0 & 1 & 1 & 3 \\ 4 & 0 & 0 & 1 & 0 & 0 & 1 \\ 5 & 0 & 1 & 1 & 0 & 0 & 3 \\ \hline & & & & & \Sigma & 11 \end{array}$ Platt number $= 11$
3	Zagreb index	$Zagreb = \sum_i \delta_i^2$ N is the total number of vertices and δ_i is the vertex degree of the ith vertex	n-Propane	$\begin{array}{c\|ccc} i & 1 & 2 & 3 \\ \hline \delta_i & 1 & 2 & 1 \\ \delta_i^2 & 1 & 4 & 1 \end{array}$ Zagreb index $= 6$
4	Balaban index	$J = \frac{M}{\mu+1}\sum_{edges}^{all}\left(\delta_i\delta_j\right)^{-0.5}$ M is the number of edges, μ denotes cyclomatic number, and δ_i and δ_j represent	Acetone	Distance matrix elements: $\begin{array}{c\|cccc\|c} & 1 & 2 & 3 & 4 & \Sigma \\ \hline 1 & 0 & 1 & 2 & 2 & 5 \\ 2 & 1 & 0 & 1 & 1 & 3 \\ 3 & 2 & 1 & 0 & 2 & 5 \\ 4 & 2 & 1 & 2 & 0 & 5 \\ \hline & & & & \Sigma & 11 \end{array}$

(continued)

Table 3.1 (continued)

Sl. No	Name of the descriptor	Mathematical formula	Sample molecule	Calculation
		the vertex degree of vertices i and j, respectively		$M = 3,\ \mu = 0$ Balaban index $= 3 \times (3 \times 15^{-0.5}) = 2.324$
5	Randić connectivity index	$\chi_R = {}^1\chi = \sum_{i=1}^{n-1}\sum_{j=i+1}^{n} a_{ij} \cdot (\delta_i \cdot \delta_j)^{-0.5}$ n is the total number of vertices, a_{ij} denotes the adjacency matrix elements, δ_i and δ_j represent the vertex degree of vertices i and j, respectively	2-Aminoacetic acid	Vertex-adjacency matrix elements: $\begin{array}{c\|ccccc\|c} & 1 & 2 & 3 & 4 & 5 & \delta \\ \hline 1 & 0 & 1 & 0 & 0 & 0 & 1 \\ 2 & 1 & 0 & 1 & 0 & 0 & 3 \\ 3 & 0 & 1 & 0 & 1 & 0 & 2 \\ 4 & 0 & 0 & 1 & 0 & 0 & 1 \\ 5 & 1 & 0 & 0 & 0 & 1 & 1 \end{array}$ Randić connectivity index $= (3^{-0.5} \times 3^{-0.5} \times 6^{-0.5} \times 2^{-0.5}) = 2.270$
6	Kier and Hall connectivity index	${}^m\chi_t = \sum_{k=1}^K \left(\prod_{i=1}^n \delta_i\right)_k^{-0.5}$ ${}^m\chi_t^v = \sum_{k=1}^K \left(\prod_{i=1}^n \delta_i^v\right)_k^{-0.5}$ k runs over the mth-order subgraphs comprising n vertices, and K corresponds to the total number of mth-order subgraphs present in the system. δ_i and δ_i^v are the vertex degree and valence vertex degree, respectively	1-Amino-2-methyl-prop-1-ene	$\begin{array}{c\|ccccc} \text{Atom} & 1 & 2 & 3 & 4 & 5 \\ \hline \delta_i & 1 & 3 & 3 & 2 & 1 \\ \delta_i^v & 1 & 4 & 3 & 3 & 1 \end{array}$ ${}^0\chi = (1^{-0.5} \times 3^{-0.5} \times 2^{-0.5} \times 1^{-0.5} \times 1^{-0.5}) = 4.284$ ${}^0\chi^v = (1^{-0.5} \times 4^{-0.5} \times 3^{-0.5} \times 3^{-0.5} \times 1^{-0.5}) = 3.655$ ${}^1\chi = (3^{-0.5} \times 6^{-0.5} \times 2^{-0.5} \times 3^{-0.5}) = 2.270$ ${}^1\chi^v = (4^{-0.5} \times 12^{-0.5} \times 9^{-0.5} \times 4^{-0.5}) = 1.622$ ${}^2\chi = (6^{-0.5} \times 6^{-0.5} \times 6^{-0.5} \times 3^{-0.5}) = 1.802$ ${}^2\chi^v = (12^{-0.5} \times 4^{-0.5} \times 36^{-0.5} \times 12^{-0.5}) = 1.244$ ${}^3\chi = (6^{-0.5} \times 6^{-0.5}) = 0.816$ ${}^3\chi^v = (36^{-0.5} \times 36^{-0.5}) = 0.333$
7	Kappa shape indices	${}^1\kappa = 2 \times \dfrac{{}^1P_{max} \times {}^1P_{min}}{({}^1P_i)^2}$; ${}^2\kappa = 2 \times \dfrac{{}^2P_{max} \times {}^2P_{min}}{({}^2P_i)^2}$; ${}^3\kappa = 4 \times \dfrac{{}^3P_{max} \times {}^3P_{min}}{({}^3P_i)^2}$ ${}^1P_i$, ${}^2P_i$ and ${}^3P_i$, respectively, denote the numbers of one, two, and three path lengths	Isobutane	$\begin{array}{c\|cccl} \text{Order} & P_i & P_{max} & P_{min} & \text{Kappa index value} \\ \hline 1 & 4 & 10 & 4 & {}^1\kappa = 2 \times \frac{10 \times 4}{4^2} = 5.0 \\ 2 & 4 & 6 & 3 & {}^2\kappa = 2 \times \frac{6 \times 3}{4^2} = 2.25 \\ 3 & 2 & 2 & 2 & {}^3\kappa = 4 \times \frac{2 \times 2}{2^2} = 4.0 \end{array}$

attempted to present the computed values of some of the indices along with their corresponding graph theoretical matrix.

Topological descriptors give highly reproducible chemical information in less time and with limited resources since they are derived from definite graph theoretical mathematical operators employing simple molecular representation. Hence, such descriptors are especially useful while dealing with a large volume of chemicals such as virtual screening study. It is to be mentioned that although topological descriptors are essentially derived from hydrogen-suppressed chemical graph theoretical formalism corresponding to two-dimensional molecular geometry, various weighting schemes can be easily incorporated in it. Sometimes topological descriptors are argued as weak since the formalism does not consider 3D features such as volume, surface area, and density which are known to depict intrinsic molecular nature. However, some of the researchers have shown the topological descriptors not to be completely devoid of spatial three-dimensional characteristics. The bonding schemes as defined in the topological formalism can be related to three-dimensional geometrical feature which is identified as 'topography' [12]. Stankevich and coworkers [14] showed a quantum chemical basis for the chi (χ) indices in terms of an energy dependence depicted by molecular electron density of conjugated hydrocarbons. The connectivity index of Randić has also been subjected to correlation with Hückel molecular orbital (HMO) parameters giving interpretation for electronic and vibrational molecular energy [15].

3.4.4 Applications

Although we have so far discussed the implication of chemical graphs toward the derivation of quantitative molecular descriptors, such graphs can be useful in other purposes too, namely canonical coding, constitutional symmetry perception, reaction graph, synthon graph, and optimal planning graph. However, considering the focus of this chapter, we shall stick to the descriptors, i.e., topological descriptors derived from chemical graphs. Like other descriptors, various topological indices represent ligand-based features. Because of algorithmic simplicity and speedier computation, topological descriptors are widely used in various issues related to chemical responses. They have suitable application in the drug designing paradigm involving virtual screening and identification of leads, lead optimization, prediction of physico-chemical property, and risk assessment of chemicals. However, in many instances, the use of topological descriptors alone might yield models with limited interpretability. Simultaneous use of thermodynamic, three-dimensional, or sometimes even one-dimensional descriptors along with topological predictors can improve the interpretative and predictive nature of a developed QSAR model [10, 11].

3.5 Three-Dimensional QSAR

3.5.1 In Silico Representation of Molecular Structure

The advancements in theoretical chemistry are aided by various computer-based applications. The implication of computer here is mainly twofold, namely analysis and storage of data. The analysis functionality enables visualization, computation of chemical information, development of models including their validation, as well as various other structure-based studies. All these different purposes are served by the use of specific software tools. The use of computer technology for solving problems in chemistry is usually known as 'computational chemistry.' It is imperative that the complex chemical analysis as well as data processing jobs performed using in silico environment is highly accurate and involves minimal laboratory resources. A brief overview of the fundamental operations performed using various computer-based applications is discussed below.

(a) Structure drawing and visualization:
     Computers provide a suitable graphical user interface for visualizing molecular structures and thereby enhancing the chemical perceptiveness of the user. The drawing of molecular structures is usually performed in a workspace supplied with chemical drawing/sketching tools such as atoms, bonds, chains, various templates for rings, amino acids, conjugated aromatic system, and fused aromatic system. The sketched chemical structure is encoded by suitable coordinates in the background which undergoes graphical conversion into images, thus allowing the user to obtain a hypothetical visualization of the drawn molecule. A particular chemical structure can be pictured in different graphical forms, namely Corey–Pauling–Koltun (CPK), stick, ball and stick, space fill, mesh, and ribbon. The ribbon representation is usually applied for macromolecules such as proteins and nucleic acids. In Fig. 3.13, we have provided different graphical visualization of the drug molecule ibuprofen. Various commercial software packages not only include different color coding for atoms and bonds for attractive and better visualization but also provide other features such as IUPAC naming, elemental analysis, and descriptor calculation. Names of some software tools for drawing of chemical structure are ChemDraw (http://www.cambridgesoft.com/Ensemble_for_Chemistry/ChemDraw/), IsisDraw (http://mdl-isis-draw.updatestar.com/), MarvinSketch (https://www.chemaxon.com/), ChemSketch (http://www.acdlabs.com/products/draw_nom/draw/chemsketch/), MedChem Designer (http://www.simulations-plus.com/Default.aspx), etc.
     The visualization functionality has a greater impact in understanding receptor–ligand interactions in structure-based modeling studies. Many commercial software packages allow three-dimensional visualization of ligand molecule as well as receptor. Such visualization is very helpful for understanding the molecular interaction taking place inside the complex biological system. It might be noted here that the visualization plane in a normal computer is two-dimensional, and hence, a three-dimensional object is displayed by using

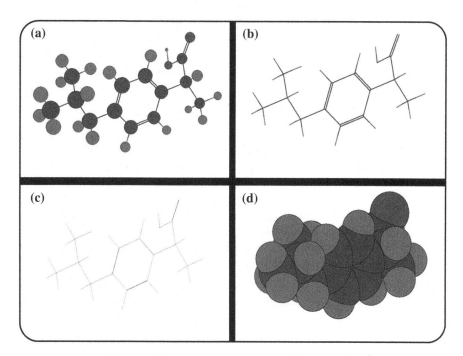

Fig. 3.13 Sample graphical visualization forms for the compound ibuprofen: **a** ball and stick, **b** stick, **c** wireframe, **d** spacefilling models

suitable modifications with respect to the axes such as change in density of color coding and change in apparent bond length (i.e., toward and away from the user). Chem3D (http://www.cambridgesoft.com/Ensemble_for_Chemistry/ ChemDraw/), Discovery Studio (http://accelrys.com/products/discovery-studio/), Sybyl (http://tripos.com/index.php), HyperChem (http://www.hyper. com/), Maestro (http://www.schrodinger.com/Maestro/), etc., are some of the commercial software platforms enabling three-dimensional view of chemical/ biological objects.

(b) Calculation and simulation:

Computer provides a stable and user-friendly environment for performing a large number of simple-to-complex computational analyses. Encoding of various formulae involving the concepts of chemistry, physics, and mathematics in software programs has enabled an in-depth study of the electronic environment of molecules at the orbital and suborbital levels. Various software programs also allow encoding of mathematical formula for the computation of chemical attributes (i.e., descriptors) useful in QSAR analysis. The mathematical formalism of any descriptor is encoded in the software platform using a suitable logical algorithm followed by the implementation of a graphical user interface such that the user can derive the computed data from simple molecular inputs. The simulation feature of computational chemistry also allows energy-based calculations as well as execution of structure-based studies, and this may be attributed to the molecular mechanics (MM) and quantum mechanics based calculations. It can be noted that computation of descriptors for QSAR studies may or may not involve energy minimization operation (if they are lower in dimension than 3D), while 3D descriptors may be derived from the computed molecular mechanical and quantum mechanical features.

(c) Data analysis and storage:
Cheminformatics studies involve generation of a significant amount of data and hence require a suitable platform for their management. QSAR analysis uses data of two major natures, the response value of chemicals and descriptors. For a large data set of chemicals, the generated data matrix becomes very large. Use of a logic-based software algorithm can easily handle such large matrix of data and gives accurate results. The determination of different validation metrics can also easily be done from computer operated software tools. Furthermore, computer allows the user for storage of chemical data in suitable formats which can be called in at any time by the user.

Figure 3.14 shows the importance of computer in combining the knowledge of chemistry along with other natural scientific disciplines, namely physics, mathematics, and biology.

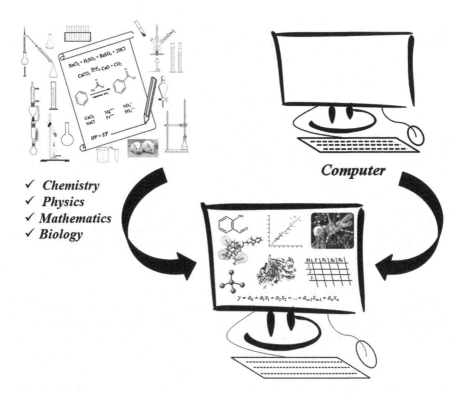

Fig. 3.14 A presentation of incorporation of chemical knowledge into computer

3.5.2 Computational Chemistry for Property Simulation

3.5.2.1 Conformational Analysis

'Conformation' can be designated as different arrangements of atoms in a molecule which is interconvertible by performing rotation about single bonds. Free rotation about sigma bond (σ) leads to a change in energy among different conformers of a molecule. According to the Newman projection, a simple molecule like butane can exist either in energetically favored staggered conformation or in unfavorable eclipsed conformation. However, numerous possible conformations can exist between these two and are known as skew conformations. The torsional strain of a molecule increases as it undergoes a rotational change from the staggered to the eclipsed conformation and become energetically unfavorable. The forces influencing conformational stability of a molecule include van der Waals force, dipole–dipole interaction and hydrogen bonding. Now, the behavioral expression of chemicals especially with respect to biological (and toxicological) response is crucially determined by the suitable arrangement of the atoms in a molecule in the three-dimensional space, and hence, conformational analysis plays an essential role in monitoring the nature of the chemical. Conformational analysis of a molecule aims in determining the minimum energy, i.e., the energetically stable form of a molecule by congregating knowledge on the flexibility of a bioactive chemical like a drug [16]. Conformational analysis has an immense importance in-silico studies such as molecular docking, screening of chemical library, and optimization of leads.

The identification of a low-energy molecular conformer is done by performing a 'search' operation using a specific algorithmic approach such as systematic search method, model-building method, random approach, distance geometry-based method, and Monte Carlo method. The principle objective is the computation of variation of torsion angle (systemic and stochastic), stochastic variation of Cartesian coordinates, stochastic variation of internuclear distances, flipping, flapping, flexing of rings or mapping of the rings onto generic shapes, etc.

3.5.2.2 Energy Minimization

Minimization of the potential energy of a molecule is essential for the determination of the stable molecular arrangement in the three-dimensional spaces. Various energy components such as stretching, bending, and torsion comprise the potential energy of a molecule, and as soon as a minimization algorithm is run in a computational platform, it immediately reaches to a minimum energy value known as 'local energy minimum,' and it could stop at that step if the used minimization method is not absolutely exhaustive. The stable conformer obtained at this stage is structurally closest to the starting molecule. Now, use of an algorithm that increase molecular strain can assist in overcoming this energy barrier and lead to the most stable conformer of a molecule termed as 'global energy minimum.' Determination of the

Fig. 3.15 Depiction of the change in potential energy of a sample molecule undergoing energy minimization operation

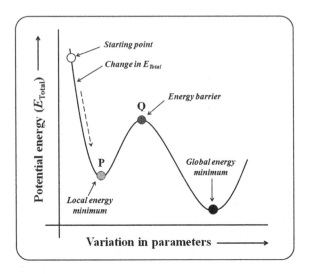

potential energy hypersurface of a stable molecule enables proper estimation of its behavior. Molecular dynamics (MD), a molecular modeling technique, allows resolving the most stable conformer of a molecule. Now, considering the biological activity of a participating molecule, it will be more interesting to put emphasis on the conformer that gives desired biological output. Such conformer may be termed as the 'bioactive conformer' and studies have depicted that it might be different from that of the 'most active conformer' which is considered as biologically potent. However, the biologically active conformer of a molecule usually stays in a zone close to the most active one. The bioactive conformation of a molecule is usually represented by the co-crystal geometry of the ligand molecule bound at a receptor site. However, the most stable conformer can be considered as the biologically active one if co-crystallized ligand is absent. Determination of global energy minimum of a ligand molecule along with its co-crystal geometry allows the user to perform comparative studies. In Fig. 3.15, the different stages of energy minimization operation have been presented.

It is to be noted that conformation analysis is used for the identification of minimum energy structures, while other simulation methods such as MD and Monte Carlo simulation lead to an assembly of states comprising of structures not at energy minima. The Monte Carlo simulation and MD methods, however, can be used as a part of the conformational search operation.

3.5.2.3 Molecular Mechanics

MM is a simulation operation which employs equations of 'classical physics' enabling the computation of various bonded attributes such as bond stretching, angle bending, and torsional energy along with other non-bonded features. MM considers the attractive and repulsive forces to control the relative positions of the nuclei of the atoms constituting a structure [17, 18]. The potential energy of a given molecule can be represented by the following simplified equation (Eq. 3.19).

$$E_{\text{Total}} = \sum E_{\text{Stretching}} + \sum E_{\text{Bend}} + \sum E_{\text{van der Waals}} + \sum E_{\text{Coulombic}} + \sum E_{\text{Torsion}}$$

(3.19)

Here, a mechanical model is hypothesized considering that spheres representing atoms are joined by mechanical springs representing covalent bonds. The energy terminologies shown in Eq. (3.19) have been formally defined in Table 3.2. The interaction and energy functionalities explained by classical physics are also termed as 'force fields.' The steric energy for a molecule is first determined employing force fields followed by the adjustment of conformational stability leading to the minimization of the steric energy. A MM operation can be executed (i) by using a commercial force field program (in a computer) as well as (ii) by exercising a molecular modeling program that assembles predefined structural fragments. While running a minimization algorithm, a molecule undergoes twisting in order to account for the steric hindrance and attractive forces, and hence, its coordinates are changed. By the use of a suitable graphical package (as available in many commercial programs), the minimization process can be recorded in the form of a video showing gradual conformational change in a structure with respect to the steric strain. MM2, MM3, MMFF, Amber, Dreiding, UFF, etc., are the examples of some force fields employed in molecular mechanical calculations.

3.5.2.4 Molecular Dynamics

This simulation technique enables mimicking the dynamic behavior of a molecule unlike other techniques such as MM that deals with static molecules at 0 K temperature, i.e., frozen state. MD technique is an in silico simulation operation allowing the prediction of time evolution of an interacting particular system [19]. This involves numerical integration operation using Newton's equation of motion followed by quantification of molecular properties in a given time frame. MD calculation can be achieved by using the following approaches.

(a) Classical mechanical formalism: Here, molecules are denoted by classical 'ball and stick' model where atoms represent soft balls while elastic sticks depict bonds. Laws of classical mechanics are employed for calculations.

(b) Quantum mechanical formalism: It was pioneered by Car and Parinello's quantum nature of chemical bond and is also known as the 'first-principles' MD simulation. The bonding in a chemical system is determined using quantum calculations, while the ions (nuclei with the inner electrons) are subjected to classical operation.

The five different conditions defining MD simulation consist of boundary condition, initial condition, force calculation, integrator/ensemble, and property

Table 3.2 Formal definition of the energy terms that are used to designate the total energy of a molecule

Force/treatment	Equation	Brief details
Torsion	$E_{\text{Torsion}} = \frac{1}{2}k_\phi[1 + \cos m(\phi + \phi_{\text{offset}})]$ where ϕ_{offset} is the ideal torsion angle relative to a staggered conformation of two atoms and k_ϕ represents the energy barrier for rotation about the torsion angle ϕ. The periodicity of rotation is denoted by 'm'	Torsional energy presents the energy required for free rotation of a sigma bond. The dihedral angle depicting the relative orientation of the atoms is the 'torsion angle.' The following figure shows the torsion angle ϕ between two sample atoms in a staggered conformation
Bond stretching	$E_{\text{Stretching}} = \frac{1}{2}k_{\text{stretch}} \times (r - r_0)^2$ where the ideal and stretched bond lengths are, respectively, denoted by r_0 and r, and k_{stretch} is a force constant giving a measure of the strength of the spring, i.e., bond	Hooke's law can be employed for the computation of bond stretching energy considering a covalent bond to be made up of a spring. However, Morse function containing complex mathematical terms also allows computation of bond stretching
Angle bending	$E_{\text{Bend}} = \frac{1}{2}k_\theta \times (\theta - \theta_0)^2$ where the ideal bond angle is denoted by θ_0 and θ is the bond angle in the bend position	The ideal bending angle is the angle formed by three consecutive atoms at their minimum energy position. Bending angle θ can be represented as follows:
van der Waals force	$E_{\text{vdW}} = \varepsilon \times \left[\left(\frac{r_{\text{min}}}{r}\right)^{12} - 2 \times \left(\frac{r_{\text{min}}}{r}\right)^6\right]$ Here, at minimum energy value ε, r_{min} presents the distance between atoms i and j, while the actual distance between the atoms is r	The van der Waals force of interaction can be represented by Lennard–Jones potential equation where the first term bearing power 6 $\{()^6\}$ represents forces of attraction and the term with 12th power $\{()^{12}\}$ denotes short-range repulsive forces involved
Coulombic force	$E_{\text{Coulombic}} = \frac{q_i \times q_j}{D \times r_{ij}}$ where q_i and q_j represent the point charges on atoms i and j, respectively, with r_{ij} being the distance between them. D denotes the dielectric constant of the medium	It measures the effect of charges between two points. The attractive or repulsive interaction between two atoms i and j separated by distance r_{ij} may be denoted as:

calculation. The method requires the input of a set of initial conditions representing initial position, particle velocity, and the interaction potential among the particles which is followed by solving a series of equations of motion for all considered particles. The force F_i acting upon the ith particle having mass m_i at the time *t* among a set of interacting particle can be denoted by the following equation based on the principles of classical mechanics.

$$F_i = m_i \frac{d^2 r_i(t)}{dt^2}$$ (3.20)

where $r_i(t)$ is the position vector of the *i*th particle and can be represented as $r_i(t) = \{x_i(t), y_i(t), z_i(t)\}$. The integration form of Newton's force equation provides the position $r_i(t + \Delta t)$ at time $(t + \Delta t)$ for the already-known positions at time *t*, and it can be mathematically presented as follows:

$$r_i(t + \Delta t) \cong 2r_i(t) - r_i(t - \Delta t) + \frac{F_i(t)}{m_i} \Delta t^2$$ (3.21)

Alternative *leapfrog, velocity* Verlet scheme, etc., can be employed for the computation of velocity. Usually, the trajectories give an infinitesimal small integration step, e.g., at the subfemtosecond scale for simulating bonds possessing light atoms to ensure the stability of the integration. Incorporation of improvement algorithms such as RESPA, SHAKE, RATTLE, and LINCS enhance the performance of MD simulation.

3.5.2.5 Quantum Mechanics

The Basic Formalism

Quantum mechanics is pioneered by Erwin Schrödinger during the study on a mathematical expression correlating motion and energy of electron [20]. The electrons were assumed to depict the wave property in Schrödinger's formalism, and hence, the equations are termed as 'wave equations,' while the series of solutions derived thereof are named 'wave functions.' Wave functions are represented as time-dependent state function for designating the nature and property of a molecular system. The basic proposition of Schrödinger wave nature of electron can be represented by the following equation.

$$H\psi = E\psi$$ (3.22)

where ψ represents the time-dependent wave function, H being the Hamiltonian operator, while E represents energy. The total potential and kinetic energy of all the particles of the molecular structure is actually denoted by the term $E\psi$. Accounting for the three-dimensional movement of electrons in space defined by *x*-, *y*- and *z*-axes, the following differential equation can be more appropriate:

$$\frac{\partial^2 \psi}{\partial x^2} + \frac{\partial^2 \psi}{\partial y^2} + \frac{\partial^2 \psi}{\partial z^2} + \frac{8\pi^2 m}{h^2}(E - V)\psi = 0$$ (3.23)

where m denotes mass, h represents Planck's constant, and the total and potential energy are represented by E and V, respectively. By the use of Laplacian operator ∇^2 for the partial differentials, the previous equation can be represented as given below:

$$\nabla^2 \psi + \frac{8\pi^2 m}{h^2}(E - V)\psi = 0 \tag{3.24}$$

The quantum mechanical principles consider the following assumptions for defining the electronic nature.

(i) Nuclei of atoms are 'motionless' considering fast motion of the electrons. This renders the nuclear energy to be separated from energy of the electrons.
(ii) An electron is characterized by its 'independent' movement which assumes an average influence of other electrons as well as the nuclei.

Considering the effects of only kinetic and potential energy terms, the Schrödinger equation can be simplified in the following equation in which the summed contribution of kinetic energy term K and potential energy term U denotes the total energy E.

$$H\psi = (K + U) \times \psi \tag{3.25}$$

Computation for H involves lengthy and complex operation, and actually for molecules containing more than 50 atoms, such treatment is not economically viable. For a simple molecule hydrogen (H_2) possessing two electron and two nuclei, the Hamiltonian operator H is given by eight terms as shown below:

$$H = -\frac{1}{2} \times \overline{V}_1^2 - \frac{1}{2} \times \overline{V}_2^2 + \frac{1}{R_1 R_2} - \frac{1}{R_1 r_1} - \frac{1}{R_1 r_2} - \frac{1}{R_2 r_1} - \frac{1}{R_2 r_2} + \frac{1}{r_1 r_2} \tag{3.26}$$

where the kinetic energy expressions for electrons 1 and 2 are represented by $\frac{1}{2}\overline{V}_1^2$ and $\frac{1}{2}\overline{V}_2^2$, respectively, while r_1 and r_2 represent the position of two electrons and the positions of their respective nuclei are denoted by R_1 and R_2.

The quantum chemical calculations can be performed by using ab initio method, density function theory (DFT)-based calculations as well as semi-empirical calculations. The quantum chemical ab initio formalism (i.e., from the beginning) attempts to furnish absolute solution to the equations characterized by high-quality accurate results using a convergent approach. However, such process becomes complex for medium-to-large-sized molecules. In order to reduce the computation burden as well as cost involved, various *less important* terms are eliminated from calculation by applying assumptions. The DFT method gives a favorable performance considering the accuracy of result, cost, and time involved, whereas calculations involving semi-empirical assumptions are reasonably fast and applicable to large chemical systems although their accuracy is lower than that of the others.

The Born–Oppenheimer Approximation

Born and Oppenheimer [21] introduced this approximation (BO approximation) to reduce the computational burden of solving time-dependent wave equation of Schrödinger. Here, the kinetic energy terms for the nuclei are neglected by assuming the nucleus to be stationery with respect to the electrons. That is, BO approximation considers the electronic motion and the nuclear motion to be separated, and hence, the wave functions of electrons are dependent on the position of nucleus, and not on its velocity. Considering the position of electrons as r_i and the position of nucleus as R_j, the following equation can be written:

$$\psi_{\text{molecule}}(r_i, R_j) = \psi_{\text{electrons}}(r_i, R_j) \cdot \psi_{\text{nuclei}}(R_j) \qquad (3.27)$$

The example of a benzene molecule may be cited here. Benzene structure consists of 42 electrons and 12 nuclei, and therefore, a partial differential eigenvalue equation of 162 variables will be obtained while dealing with the energy and molecular wave functions using Schrödinger equation. Such difficulty in calculation can be eased by the employment of BO approximation.

The Hartree–Fock Approximation

The Hartree–Fock (HF) approximation is attributed to the decisive contribution of Hartree [22] and Fock [23]. This operation is also known as the 'self-consistent field (SCF)' method. Hartree formulated an approximation for a many-electron wave function system by using the products of single particle functions which can be depicted as follows:

$$\psi(r_1, r_2, \ldots, r_n) = \phi_1(r_1) \times \phi_2(r_2) \times \cdots \times \phi_n(r_n) \qquad (3.28)$$

where $\psi(r_1, r_2, \ldots, r_n)$ is a many-electron wave function with r_i denoting coordinates and spins of the particles. Each of the functions $\phi_i(r_i)$ corresponds to one-electron, i.e., one-particle Schrödinger equation and for a total number of N electrons, Hartree defined it as follows:

$$\left[-\frac{1}{2}\Delta + v(r) + \sum_{j=1, j \neq 1}^{N} \int \frac{|\phi_j(r')|^2}{|r - r'|} \, \mathrm{d}r' \right] \phi_i(r) = E_i \phi_i(r) \qquad (3.29)$$

where $v(r)$ corresponds to the nuclear charge measure Z which can be defined as: $v(r) = -Z/r$. Here, an electron is assumed under the 'SCF' at the ith state that is determined by all electrons but the ith one. The HF approximation experiences violation of exclusion principle due to the non-orthogonal nature of the functions $\phi_i(r)$. 'Anti-symmetrized modification,' 'Fermi statistics inclusion,' 'configuration interaction (CI),' etc., represent the modifications as well as extensions incorporated into the HF formalism. A HF equation is iteratively solved using suitable algorithmic platform in computer.

Density Function Theory (DFT)

The DFT [24] considers the electronic motions as 'uncorrelated,' while a local approximation of the free electrons can be used to represent the kinetic energy. Thomas and Fermi provided the initial concept of DFT which can be furnished by the following integral: $n(r) = N \int dr_2 \ldots \int dr_N \psi * (r, r_2, \ldots, r_N) \times \psi(r, r_2, \ldots, r_N)$, where the density of the electron is denoted by $n(r)$. The DFT formalism evolved and went under refinement while addressing drawbacks of the HF approximation. The theorem for DFT was provided by Hohenberg and Kohn which was later simplified by Levy. For a system with N electrons moving in an external potential $V_{ext}(r)$, the Hamiltonian operator H can be represented as follows:

$$H = T + V_{ee} + \sum_{i=1}^{N} V_{ext}(r_i) \qquad (3.30)$$

where the kinetic energy and the electron–electron interaction operators are, respectively, represented by T and V_{ee}. If ψ_{GS} represents the wave function and $n_{GS}(r)$ the density, the ground-state energy E_{GS} can be mathematically formulated as follows:

$$\begin{aligned} E_{GS} &= \int dr V_{ext}(r) n_{GS}(r) + \langle \psi_{GS}|T + V_{ee}|\psi_{GS}\rangle \\ &= \int dr V_{ext}(r) n_{GS}(r) + F[n_{GS}] \end{aligned} \qquad (3.31)$$

where $V_{ext}(r)$ represents the external potential, while $F[n]$ denotes a density functional such that it is not dependent to any specific system or external potential. Kohn and Sham used local density (LD) approximation to the limiting case of a slowly varying density and provided a means for solving Schrödinger equation for a fictitious system of non-interacting particles (Eq. 3.32).

$$E_{xc}^{LD} = \int dr\ n(r)\ \varepsilon_{xc}[n(r)] \qquad (3.32)$$

where $\varepsilon_{xc}[n]$ represents the exchange and correlation energy per particle of a homogeneous electron gas possessing density n.

The problem of non-locality of single particle exchange potential in HF approximation is overcome by Kohn–Sham local density approximation. DFT enables computation of vibrational frequencies, atomization energies, ionization energies, electric and magnetic properties, reaction paths, etc.

Semi-empirical Analysis

Semi-empirical analysis assumption of the quantum chemical analysis employs integral approximations and parameterizations aiming to reduce the complexity of solving Schrödinger wave equation. Such hypothesis can operate for larger molecular systems although with 'less accurate' computation outcomes. The semi-empirical calculations start with the ab initio method and soon after speeds up the computation by avoiding various less important terms and features thereof. However, in order to compensate the assumptive errors, such methods use empirical parameters with calibration against reliable theoretical or experimental data and hence are termed as the 'semi-empirical' techniques [25].

HMO method for the generation of molecular orbital values of unsaturated molecules using π-electronic formalism can be identified as the oldest form of semi-empirical approach. However, the Hückel-type methods use one-electron integrals and are also non-iterative in nature. The Pariser–Parr–Pople ideology depicts the electronic spectra of unsaturated molecules using anti-symmetrized products of quantitative atomic orbital integrals bearing the core Hamiltonian. This method institutes the idea of zero differential overlap along with the charged sphere form of atomic orbitals. People showed that the neglect of differential overlap in electron interaction integral without further adjustments is not constant to simple transformation of the atomic orbital basis set such as the s and p orbital replacement by hybrids or the rotation of axes. Modern approaches based on the semi-empirical formalism employ methods of neglect of diatomic differential overlap (NDDO), intermediate neglect of differential overlap (INDO), with the complete neglect of differential overlap (CNDO) [25]. Some of the methods have been briefly discussed in Table 3.3.

It will be noteworthy to mention that among different platforms allowing molecular simulation operations, Gaussian software (http://www.gaussian.com/) allows scrupulous theoretical computation involving the ab initio formalism (HF, MP2, etc.), density functional theory (HFB, PW91, PBE, G96, LYP, VWN5, etc.), semi-empirical techniques (AM1, MNDO, PM3, PM6, etc.), MM (Amber, Dreiding, UFF), and other hybrid methods (G1, G2, G2MP2, G3, G3B3, G4, G4MP2, MPW1PW91, B2PLYP, B3LYP, etc.). The Gaussian software additionally characterizes the wave functions using various 'basis sets,' namely STO-3G, 3-21G, 6-21G, 4-31G, and 6-31G. This software is also the oldest one in this genre and is accredited to John Pople and his research group [26] who released the first version in the year 1970 (*Gaussian70*).

3.5.3 Examples of 3D-QSAR

The aim of any 3D-QSAR is to establish the relationship between biological activity and spatially localized steric, electrostatic, lipophilic, and hydrogen-bonding properties of chemicals. The 3D-QSAR approaches are computationally more

Table 3.3 A representative view of different schemes which have been implemented in semi-empirical/self-consistent quantum chemical calculations

Sl. No.	Abbreviated name	Full form of the technique	Brief note
1	LCAOSCF	LCAO self-consistent function	Provides self-consistent function approximation using LCAO method. Here, energy minimization is facilitated by the coefficient of the orbitals
2	CNDO	Complete neglect of differential overlap	CNDO and NDDO are the simplified forms of LCAOSCF using the approximation of neglecting differential overlap. CNDO does not consider any differential overlap in all the basis sets. That is, here a product of two different atomic orbitals corresponding to a specific electron is always 'neglected' in electron interaction integrals
3	NDDO	Neglect of diatomic differential overlap	It corresponds to the product of pairs of atomic orbitals of different atoms that have been neglected in certain electron repulsion integral
4	INDO	Intermediate neglect of differential overlap	It represents the neglect of the differential overlap in the integral of all electron interaction except those using one center only, i.e., the retention of one-center product of different atomic orbital in only one-center integral. It is of an intermediate complexity between CNDO and NDDO methods
5	MINDO	Modified intermediate neglect of differential overlap	It considers a common value in order to represent the two-center electron repulsion integral between the atomic orbitals of a chosen atomic pair
6	MNDO	Modified neglect of diatomic overlap	Here, the approximation has been applied to the closed shell molecules and their valence electrons which are assumed to move in a constant core field composed of the nuclei and inner shell electrons
7	AM1	Austin Model 1	An approximation of NDDO that uses nuclear–nuclear core repulsion function (CRF) for approximation of two electron integrals to mimic the van der Waals interaction
8	PM3	Parametric Method 3	It uses a Hamiltonian operator like AM1, but institutes a different parameterization strategy involving a large number of molecular properties. H-bonds are well assessed although non-physical hydrogen–hydrogen attraction causes trouble

expensive than the 2D-QSAR approaches. The 3D-QSAR techniques are broadly divided into two classes based on the alignment strategy of the studied molecules.

1. **Alignment-based techniques**:

   - Comparative molecular field analysis (CoMFA),
   - Self-organizing molecular field analysis (SOMFA),
   - Comparative molecular similarity indices analysis (CoMSIA),
   - Receptor surface analysis (RSA), and
   - Molecular shape analysis (MSA).

2. **Alignment-independent techniques:**

   - Comparative molecular moment analysis (CoMMA),
   - Weighted holistic invariant molecular (WHIM) descriptor analysis,
   - VolSurf,
   - Compass,
   - Comparative spectral analysis (CoSA), and
   - Grid-independent descriptors (GRIND)

In the present chapter, we have discussed the most commonly and successfully used methods only.

3.5.3.1 CoMFA

Perception of CoMFA

The CoMFA is a molecular field-based, alignment-dependent, ligand-based 3D-QSAR method which generates a quantitative relationship of molecular structures and its biological response [27, 28]. This method considers ligand properties such as steric and electrostatic energies, and resulting favorable and unfavorable receptor–ligand interaction. In CoMFA, all aligned ligands are placed in an energy grid and by placing an appropriate probe at each lattice point, energies are computed. The resultant energy computed at each unit point corresponds to electrostatic (Coulombic) and steric (van der Waals) properties. These calculated values serve as descriptors which are then correlated with biological responses employing linear regression methods such as partial least squares (PLS).

Formalism of CoMFA

The methodology of the CoMFA is described below:

   I. First, a set of ligands known to bind in the same binding mode and binding pocket of a receptor are taken and their structures are drawn.
   II. Then, energy minimization is carried out and the bioactive conformation of each molecule is generated.

III. Thereafter, all the molecules are superimposed employing either manual or automated methods in a manner defined by the supposed mode of interaction with the receptor.

IV. Thereafter, the overlaid compounds are positioned in the center of a lattice grid with a spacing of 2 Å.

V. The steric and electrostatic field intensities are calculated in the 3D space around the molecules with different probe groups positioned at all intersections of the lattice. Computation of the steric field follows the Lennard–Jones equation, and computation of electrostatic field follows the Coulombic interaction equation.

VI. The interaction energy (descriptors) is correlated with the biological response employing the PLS tool.

VII. Interactive graphics consisting of colored contour plots are generated for the easy interpretation of the results.

The CoMFA methodology is schematically depicted in Fig. 3.16.

Factors Responsible for the Performance of CoMFA

There are miscellaneous factors [28] concerned for quality of the constructed CoMFA model and these are explained in Table 3.4.

Fig. 3.16 Flowchart of basic steps of the CoMFA methodology

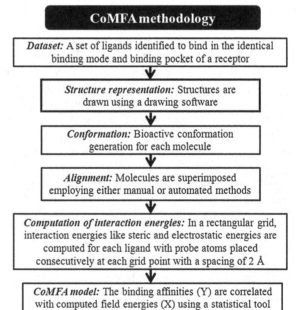

Table 3.4 Factors governing the quality of a CoMFA model

Factors	Significance
Biological data	• All molecules should belong to a congeneric series • Compounds should possess the same mechanism of action and same or at least equivalent binding mode • The biological responses of molecules should correlate to their binding affinity, and their specified biological responses should be assessable • Experimental responses should be measured employing standardized and uniform protocols and preferably from a defined endpoint and single laboratory • The activity values of all compounds should be in same units of measurement (binding/functional/IC_{50}/K_i) • The ranges of biological responses should be as large as possible, keeping the mode of action identical
Optimization of 3D structure	*Molecular mechanics*: It does not explicitly consider the electronic motion, and therefore, they are fast, accurate, and can be employed for large molecules *Quantum mechanics* or ab initio: It takes into account the 3D-distribution of electrons around the nuclei and thus is extremely precise. These methods are time-consuming, computationally intensive, and they cannot handle large molecules *Semi-empirical*: Semi-empirical quantum methods attempt to address two restrictions, namely slow speed and unsuitability for large molecules of the quantum mechanical calculations by omitting certain integrals based on experimental data, such as ionization energies of atoms, or dipole moments of molecules. Modern semi-empirical models are based on the Neglect of Diatomic Differential Overlap (NDDO) methods such as MNDO, AM1, PM3, and PDDG/PM3
Conformational search analysis	• Systematic search (or grid search) • Monte Carlo • Random search • Molecular dynamics • Simulated annealing • Distance geometry algorithm • Genetic and evolutionary algorithms
Determination of bioactive conformations	*X-ray crystallography*: The precise 3D structure of the macromolecules can be obtained from the X-ray crystal structure. Drug–receptor complexes generated by this method offer the exact information *NMR spectroscopy*: The 3D structural data are obtained in solution and it is a method of selection when the molecule cannot be crystallized through experimental ways, as in case of the membrane bound receptors or receptors which have not yet been isolated due to stability, resolution, or other issues
Alignment of molecules	• Atom overlapping-based superimposition • Binding sites-based superimposition • Field/pseudofield-based superimposition • Pharmacophore-based superimposition • Multiple conformer-based superimposition

(continued)

Table 3.4 (continued)

Factors	Significance
Computation of molecular interaction energy fields	• The standard size of the grid spacing is 2 Å. The grid spacing is inversely proportional to the accuracy of calculations. As the grid spacing decreases to 1 Å or less, the calculations become more extensive • The distinctive size of the grid box is 3–4 Å larger than the union surface of the overlaid compounds. As the electrostatic/Coulombic interactions are long-range in nature, a larger grid box may be required. It is true for steric/van der Waals interactions also • The interaction energies are computed using probes. The probe may be a small molecule like water, or a chemical fragment such as a methyl group. The electrostatic energies are calculated with H^+ probe, whereas a sp^3 hybridized carbon atom with an effective radius of 1.53 Å and a +1.0 charge is used as probe for including the steric energies • In CoMFA, the standard Lennard–Jones function is utilized to model the van der Waals interactions, whereas electrostatic interactions are determined by the Coulomb's law

Display and Interpretation of Results

The results are displayed for a CoMFA model by two ways:

(a) **Coefficient contour plots**: It portrays vital regions in space around the compounds where specific structural modifications appreciably vary with the response. In CoMFA, two types of contours are shown for each interaction energy field: (i) the positive and (ii) negative contours which are depicted by some specific colors.

(b) **Plots from PLS models**: Two types of plots are generally created: (i) score plots and (ii) loading/weight plots. The score plots between biological response (Y-scores) and latent variables (X-scores) show relationships between the activity and the structures, whereas plots of latent variables (X-scores) display the similarity/dissimilarity between the molecules, and their clustering predispositions.

Advantages and Drawbacks of CoMFA

The CoMFA has the ability to design of new ligands in the structure–activity correlation problems. Along with a good number of advantages, a CoMFA model is not free from limitations also (Table 3.5).

Table 3.5 Commonly encountered advantages and limitations of the CoMFA model

Advantages	Physicochemical features such as steric and electrostatic forces involved in ligand–receptor interactions
	Applicable to any series of molecules for which alignable models can be constructed and whose desired property is believed to result from alignment-dependent non-covalent molecular interactions
	Each CoMFA parameter represents the interaction energy of an entire ligand, not just the interaction of a more or less randomly selected substructure of the ligand
	The only inputs needed are models of all the molecules, their lattice description, and usually, an explicit 'alignment rule'
	Interpretation through the 'coefficient contour' map
Limitations	Consideration of too many variables such as overall orientation, lattice placement, step size, and probe atom type
	Appropriate only for in vitro data
	Hydrophobicity is not well-quantified
	Low signal-to-noise ratio due to many ineffectual field variables
	Improbability in choice of molecules and variables
	Fragmented contour maps with variable selection procedures
	Flaws in potential energy functions
	Cutoff limits are utilized

3.5.3.2 CoMSIA

Idea of CoMSIA

The CoMSIA is a ligand-based, alignment-dependent and linear 3D-QSAR method [29]. The major difference between CoMFA and CoMSIA is that molecular similarity is considered in case of CoMSIA. The CoMFA concentrates on the alignment of compounds and may lead to fault in alignment sensitivity and interpretation of electrostatic and steric potential. To overcome the problem, Gaussian potentials are utilized in CoMSIA fields. The usual energy grid box is created and similar probes are positioned throughout the grid lattice. In addition, solvent reliant molecular entropic (hydrophobicity) term is also included in the CoMSIA. To analyze the property of molecules, a common probe is placed and similarity at each grid point is computed. In CoMSIA, five different similarity fields are calculated at regular-spaced grid points for the aligned molecules.

- Steric,
- Electrostatic,
- Hydrophobic,
- Hydrogen bond donor, and
- Hydrogen bond acceptor.

Methodology of CoMSIA

The general formalism of the CoMSIA technique is illustrated below:

(a) Conformer generation is performed for the studied molecules.
(b) Energy minimization of the molecules is performed and then partial atomic charges of the molecules are calculated.
(c) After that, the training set molecules are aligned based on the points of alignment of the most active compound (template molecule).
(d) Thereafter, molecular interactions based on the five physicochemical properties are calculated using a common probe atom with 1 Å radius (can be extended by 2.0 Å in all directions), charge +1, hydrophobicity +1, hydrogen bond donor and acceptor properties +1.
(e) Subsequently, the PLS approach is employed to derive the 3D-QSAR models using the correlation between similarity factors and biological response.
(f) The results are illustrated in the form of contour maps which differentiate the favorable and unfavorable regions for the five different interaction fields.

The basic steps involved in the CoMSIA methodology are illustrated in Fig. 3.17.

Fig. 3.17 The entire formalism of the CoMSIA technique

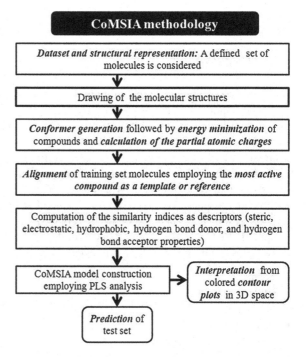

Advantages of CoMSIA

The CoMSIA technique provides following unique advantages:

- The 'Gaussian distribution of similarity indices' overcomes the unanticipated changes in grid-based probe–atom interactions.
- The choice of similarity probe includes steric and electrostatic potential fields as well as hydrogen bonding and hydrophobic fields.
- The effect of the solvent entropic provisions can also be included by employing a hydrophobic probe.
- The CoMSIA contours indicate those areas within the region occupied by the ligands that 'favor' or 'oppose' the occurrence of a group with a particular physicochemical property or response.

3.5.3.3 MSA

Concept of the MSA

The MSA technique is employed to identify the biologically relevant conformation without knowledge of the receptor geometry [30]. The MSA is an alignment-dependent approach which incorporates conformational flexibility and shape data into the 3D-QSAR. Multiple conformations of each molecule can be generated using the conformational search method. A conformer of the most active compound selected as a shape reference compound to which all the structures in the study compounds can be aligned through pairwise superpositioning. This alignment procedure looks at molecules as points and lines and uses the techniques of graph theory to identify patterns. It finds the largest subset of atoms in the shape reference compound that is shared by all the structures in the study table and uses this subset for alignment. A rigid fit of atom pairings is performed to superimpose each structure so that it overlays the shape reference compound.

Methodology of the MSA

The MSA is an iterative process, in which the molecular shape similarities and other descriptors are employed to generate a QSAR model with best possible statistical significance. The process consists of seven fundamental steps as mentioned below:

1. The first step is the generation of conformers and energy minimization of each structure to be investigated.
2. The next step generates a structure that corresponds to the structure present in the rate-limiting step for the biological action. The step is known as '*hypothesizing an active conformer.*'
3. Then, a shape reference compound is identified which is used when shape descriptors are calculated for the study matrix. Generally, the most active compound is considered as the reference compound.

Table 3.6 Commonly employed MSA descriptors

Descriptors	Definitions
DIFFV (difference volume)	Difference between the volume of the individual molecule and the volume of the shape reference compound
Fo (common overlap volume ratio)	Common overlap steric volume descriptor divided by the volume of the individual molecule
NCOSV (non-common overlap steric volume)	Volume of the individual molecule and the common overlap steric volume
ShapeRMS (rms to shape reference)	This is root mean square (rms) deviation between the individual molecule and the shape reference compound
COSV (common overlap steric volume)	Common volume between each individual molecule and the molecule selected as the reference compound
SRVol (volume of shape reference compound)	This is the volume of the shape reference compound

4. The fourth step in MSA is to execute pairwise molecular superpositions to find out what and how atoms of data set compounds are equivalent to atoms in the shape reference compound.
5. The fifth step is measuring molecular shape commonality to compare the properties that two molecules have in common.
6. In this step, the researcher can also add other molecular properties to the descriptor matrix by calculating spatial, electronic, thermodynamic descriptors, etc.
7. The final step is the construction of a QSAR equation by the application of genetic function approximation (GFA) or genetic partial least squares (G/PLS) methods.

MSA Descriptors

The MSA descriptors are used to determine the molecular shape commonality. Most commonly used MSA descriptors are incorporated in Table 3.6.

3.5.3.4 RSA

Concept of the RSA

The RSA [31] is a helpful tool when the 3D structure of the receptor is unidentified, since one can construct a hypothetical model of the receptor site using RSA. RSA varies from pharmacophore models in that the RSA approach tries to capture necessary information about the receptor, while the pharmacophore captures information about the commonality of compounds that bind to a receptor. The model embodies essential information about the hypothetical receptor site as a 3D surface with associated properties such as follows:

• Hydrophobicity,
• Partial charge,

- Electrostatic (ELE) potential,
- Van der Waals (VDW) potential, and
- Hydrogen bonding propensity.

Receptor surface models provide compact quantitative descriptors which capture 3D information of interaction energies in the form of steric and electrostatic fields at each surface point.

Methodology of the RSA

The fundamental steps of RSA are as follows:

1. First, conformers of the compounds are generated and energy minimization is performed
2. Then, compounds are superimposed in their bioactive conformation
3. Thereafter, a receptor-complementary surface is created employing shape fields which encloses a volume common to all the aligned compounds and which represents their collective molecular shape.
4. Then, assumed chemical properties of the receptor at every surface point are calculated.
5. Finally, CoMFA models are constructed by GFA or G/PLS correlating the surface properties with the biological response of the studied molecules.

RSA Descriptors

The RSA descriptors signify the energy of interactions between each point on the receptor surface and each model. The frequently used RSA descriptors are illustrated in Table 3.7.

Miscellaneous

Relatively less used 3D-QSAR methods are discussed in Table 3.8 [32]. For a better understanding of readers, these discussed methods are classified into two sections, alignment-based and alignment-independent models.

Table 3.7 A complete list of RSA descriptors

Descriptors	Definitions
IntraEnergy	Molecular internal energy inside the receptor
InterEleEnergy	Non-bond electrostatic energy between a molecule and the receptor
InterVDWEnergy	Non-bond van der Waals energy between a molecule and the receptor
InterEnergy	Total non-bond energy between a molecule and the receptor
MinIntraEnergy	Molecular internal energy minimized without receptor
StrainEnergy	Molecular strain energy within the receptor and the molecule minimized without the receptor model

Table 3.8 A bird's-eye view idea on miscellaneous 3D-QSAR methods

3D-QSAR methods	Concept/explanation
Alignment-based 3D-QSAR model	
Adaptation of the fields for molecular comparison (AFMoC)	The AFMoC is a 3D-QSAR method that considers fields derived from the protein environments and known as a 'reverse' CoMFA approach or 'Inverted CoMFA' derived from the potential scoring function. The protein-specific potential fields are generated into binding sites, which are employed for the prediction of binding affinity
Genetically evolved receptor modeling (GERM)	The GERM is helpful for developing 3D models of macromolecular binding sites in the nonexistence of experimental structure such as X-ray crystallography and NMR spectroscopy or homology-modeled structure of the target receptor. The key constraint for GERM is that all the aligned conformers should be enclosed into the receptor active site, allocating them as a shell of atoms. The allocated shells of atoms are considered an explicit set atom (aliphatic H, aliphatic C, polar H, etc.) and matched at the receptor active site analogous to those originate in the receptor active site. The drawback of the GERM methodology is that it considers only a single conformation of each ligand in the training set as well as its single orientation in the binding site
Hint interaction field analysis (HIFA)	The HIFA is a newly developed approach employing calculation of empirical hydrophobic interactions. Due to the introduction of hydrophobicity calculation in CoMFA, the predicative capability of this method has enhanced. It computes key hydrophobic features which are atom-based analogs of the fragment constant
Molecular quantum similarity measures (MQSM)	The MQSM is defined by the vectors of the electronic density function as descriptors, which represent the similarity using molecular quantum similarity measures. The MQSM is optimized by translating and rotating molecular pairs so as to make the most of the overlap of their molecular electronic density. The methodology is based on the quantification of the similarity between two molecules using the first order density functions of both studied systems
Self-organizing molecular field analysis (SOMFA)	The SOMFA technique has similarities to the molecular similarity analysis and CoMFA. It has also similarities with the hypothetical active site lattice (HASL) method. The 'mean centered activity' is crucial in SOMFA
Voronoi field analysis (VFA)	In the VFA technique, Voronoi field variables are assigned to each of the Voronoi polyhedra created by dividing the superimposed molecular space but not to each of the lattice points like CoMFA

(continued)

Table 3.8 (continued)

3D-QSAR methods	Concept/explanation
Alignment-independent QSAR model	
Comparative molecular moment analysis (CoMMA)	The CoMMA technique addresses second-order moments of the shape, mass distribution and charge distributions. The moments relate to center of the mass and center of the dipole. The CoMMA descriptors comprise principal moments of inertia, magnitudes of dipole moment and principal quadrupole moment. Descriptors relating charge to mass distributions are defined, i.e., magnitudes of projections of dipole upon principal moments of inertia and displacement between center of mass and center of dipole
Comparative spectral analysis (CoSA)	The CoSA has employed molecular spectroscopy techniques for the identification of the 3D molecular descriptors of chemicals. The molecular spectra are employed to predict biological activity of the 3D structures. The spectroscopic method generally comprises following techniques: proton (^{1}H)-NMR, carbon (^{13}C)-NMR, IR, and mass spectrometry
Compass	Compass automatically chooses conformations and alignments of molecules. In this approach, each molecule is represented by a different set of feature values. Three types of features, steric, hydrogen-bond donor and acceptor features, are used in the Compass approach. Steric distances are computed from the sampling points to the adjacent atom. Donor and acceptor feature values are calculated as the distance from the sampling points scattered near the surface of the molecules to the nearest hydrogen bond donor and acceptor groups, respectively
VolSurf	The VolSurf approach depends on probing the grid around the molecule with specific probes. The resulting lattice boxes are employed to calculate the descriptors relying on volumes or surfaces of 3D contours, defined by the same value of the probe molecule interaction energy. By using various probes and cutoff values of the energy, different molecular properties can be quantified
Weighted holistic invariant molecular (WHIM) descriptor analysis	WHIM descriptors offer the invariant information by utilizing the principal component analysis (PCA) on the constructed coordinates (Cartesian coordinates around x-, y-, z-axes) of the atoms constituting the molecule. This converts the molecule into the space that captures the most variance. In this space, numerous statistics are computed and they serve as directional descriptors, including variance, proportions, symmetry, and kurtosis

3.6 Conclusion

The classical approaches of Free–Wilson and Hansch analyses constitute the fundamental basis of QSAR/QSPR modeling. These classical techniques were based on ligand-based design strategies and suffer from various drawbacks. However, such methods provided a conceptual platform for predictive modeling analysis, and even today suitably modified versions of these techniques are used by researchers. With the advancement of knowledge in chemistry, various dimensional perspectives provided theoretical analysis of molecular features following ligand as well as structure-based formalisms. The use of computer as an abstract platform of performing various QSAR/QSPR operations has undoubtedly speeded up the research findings by allowing accurate and reliable molecular modeling analysis and data treatment.

References

1. Free SM, Wilson JW (1964) A mathematical contribution to structure-activity studies. J Med Chem 7:395–399
2. Beasley JG, Purcell WP (1969) An example of successful prediction of cholinesterase inhibitory potency from regression analysis. Biochim Biophys Acta 178:175–176
3. Purcell WP, Bass GE, Clayton JM (1973) Strategy of drug design: a guide to biological activity. Wiley, New York
4. Fujita T, Ban T (1971) Structure-activity study of phenethylamines as substrates of biosynthetic enzymes of sympathetic transmitters. J Med Chem 14:148–152
5. Hammett LP (1935) Some relations between reaction rates and equilibrium constants. Chem Rev 17:125–136
6. Hansch C, Fujita T (1964) ρ-σ-π Analysis. A method for the correlation of biological activity and chemical structure. J Am Chem Soc 86:1616–1626
7. Kubinyi H (1993) QSAR: Hansch analysis and related approaches. In: Mannhold R, Krogsgaard-Larsen P, Timmerman H (eds) Methods and principles in medicinal chemistry. VCH, Weinheim
8. Balaban AT (1985) Applications of graph theory in chemistry. J Chem Inf Comput Sci 25:334–343
9. Trinajstić N, Gutman I (2002) Mathematical Chemistry. Croat Chem Acta 75:329–356
10. Roy K (2004) Topological descriptors in drug design and modeling studies. Mol Diver 8:321–323
11. Todeschini R, Consonn V (2009) Molecular descriptors for chemoinformatics. Wiley-VCH, New York
12. Kier LB, Hall LH (1986) Molecular connectivity in structure-activity analysis. Wiley, New York
13. Basak SC, Gute BD, Grunwald GD (1997) Use of topostructural, topochemical, and geometric parameters in the prediction of vapor pressure: a hierarchical QSAR approach. J Chem Inf Comput Sci 37:651–655
14. Stankevich IV, Skovortsova MI, Zefirov NS (1995) On a quantum chemical interpretation of molecular connectivity indices for conjugated hydrocarbons. J Mol Struct (Theochem) 342:173–179

15. Gálvez J (1998) On a topological interpretation of electronic and vibrational molecular energies. J Mol Struct (Theochem) 429:255–264
16. Mazzanti A, Casarini D (2012) Recent trends in conformational analysis. WIREs Comput Mol Sci 2:613–641
17. Hendrickson JB (1961) Molecular geometry. I. Machine computation of the common rings. J Am Chem Soc 83:4537–4547
18. Allinger NL (1977) Conformational analysis. 130. MM2. A hydrocarbon force field utilizing V1 and V2 torsional terms. J Am Chem Soc 99:8127–8134
19. Petrenko R, Jaroszaw M (2010) Molecular dynamics. Encyclopedia of life sciences (ELS). Wiley, Chichester
20. Schrödinger E (1926) Quantisierung als Eigenwertproblem (Erste Mitteilung). Ann Phys 79:361–376
21. Born M, Oppenheimer R (1927) Zur quantentheorie der molekeln. Ann Phys 389:457–484
22. Hartree DR (1928) The wave mechanics of an atom with a non-Coulomb central field. I. Theory and methods. Proc Cambridge Philos Soc 24:89–110
23. Fock V (1930) Näherungsmethode zur Lösung des Quantenmechanischen Mehrkörperproblems. Z Phys 61:126–148
24. Jones RO, Gunnarsson O (1989) The density functional formalism, its applications and prospects. Rev Mod Phys 61:689–746
25. Thiel W (2014) Semiempirical quantum–chemical methods. WIREs Comput Mol Sci 4:145–157
26. Pople JA, Santry DP, Segal GA (1965) Approximate self consistent molecular orbital theory. I. Invariant procedures. J Chem Phys 43:S129–S135
27. Cramer RD III, Patterson DE, Bunce JD (1988) Comparative molecular field analysis (CoMFA). I. Effect of shape on binding of steroids to carrier proteins. J Am Chem Soc 110:5959–5967
28. Norinder U (1998) Recent progress in CoMFA methodology and related techniques. In: Kubinyi H, Folkers G, Martin YC (eds) 3D QSAR in drug design—recent advances, vol 3. Kluwer Academic Publishers, New York, pp 24–39
29. Klebe G, Abraham U, Mietzner T (1994) Molecular similarity indices in a comparative analysis (CoMSIA) of drug molecules to correlate and predict their biological activity. J Med Chem 37:4130–4146
30. Hopfinger AJ, Tokarski JS (1997) Three-dimensional quantitative structure-activity relationship analysis. In: Charifson PS (ed) Practical application of computer-aided drug design. Marcel Dekker Inc., New York, pp 105–164
31. Hahn M (1995) Receptor surface models. 1. Definition and construction. J Med Chem 38 (12):2080–2090
32. Verma J, Khedkar VM, Coutinho EC (2010) 3D-QSAR in drug design—a review. Curr Top Med Chem 10(1):95–115

Chapter 4
Newer Directions in QSAR/QSPR

Abstract The QSAR/QSPR technique is now a widely practiced tool in chemical research both in the industry and academia. Because of the enormous potential applications of predictive modeling analysis, various newer methods have recently been developed to improve the usefulness and applicability of QSAR techniques. Binary QSAR, hologram QSAR (HQSAR), group-based QSAR (G-QSAR), multivariate image analysis (MIA)-based QSAR (MIA-QSAR), etc., are some of the new approaches in the realm of QSAR formalisms. Furthermore, QSAR techniques are also employed in various newer research areas in addition to the conventional drug design and predictive toxicology paradigm. QSAR models have been observed to be fruitful in modeling various property endpoints in the field of material informatics. In addition to that, predictive modeling of properties and/or toxicities of nanoparticles (NPs), cosmetics, peptides, ionic liquids, phytochemicals, etc., also represents the emerging application areas of the QSAR technique. This present chapter gives an overview of both the new methods and new application areas of QSAR studies.

Keywords Binary QSAR · Cosmetics · G-QSAR · HQSAR · Inverse QSAR · MIA-QSAR · Mixture toxicity · Nanomaterials · Peptides · Phytochemistry

4.1 Introduction

The QSAR/QSPR modeling technique provides an opportunity for the rational design of chemicals. With the availability of various response data (activity/property/toxicity) of diverse chemical compounds, such modeling approach has been employed to monitor different scientific issues related to behavioral manifestations of chemicals. Historically, the beginning of the QSAR formalism took place in modeling various toxicological endpoints, while later it was profusely used in the field of physical organic chemistry addressing the property data. The pharmacological activity data have also been modeled to a significant extent by the

© The Author(s) 2015
K. Roy et al., *A Primer on QSAR/QSPR Modeling*,
SpringerBriefs in Molecular Science, DOI 10.1007/978-3-319-17281-1_4

researchers in this field. With the passage of time, QSAR has become a well practiced study to predict and/or fine-tune physicochemical properties, activity potential, and toxicological hazard of chemicals in relation with structures. Other than just providing a logical correlation of data, QSAR also enables nurturing the mechanistic basis involved in a biological, toxicological, or physicochemical process. This feature has enabled the use and application of QSAR/QSPR techniques to a greater extent in different areas of chemical sciences. Presently, various international regulatory authorities propose the use of QSAR techniques as a suitable alternative strategy to in vivo biological (and toxicological) experimentation. Specific regulations and 'expert systems' have also been presently developed by various countries. Hence, it may be observed that the QSAR analysis is presently used by a broader part of the scientific community, and with the aim of enhancing the performance of this formalism, various newer techniques have evolved. The concept of specialized molecular fragments has been implemented in relatively new techniques, namely hologram QSAR (HQSAR) [1, 2] and group-based QSAR (G-QSAR) [3]. The multivariate image analysis (MIA)-based QSAR (MIA-QSAR) [4] method also presents a new QSAR method based on the attributes of two-dimensional image of molecular structures. Among other new QSAR techniques [5], we can mention about LQTA–QSAR, eQSAR, and novel approaches such as FB-QSAR, FS-QSAR, SOM-QSAR, QUASAR (5D-QSAR), and 6D-QSAR. Along with the development of various techniques, the newer application opportunities of the QSAR technique in chemical research have also come into the light. QSAR actually presents a ligand-based approach of exploring structural features of chemicals responsible for an activity or property. However, it has been observed to be very fruitful in the paradigm of 'lead designing' when used in combination with structure-based approaches. With the acceptance of different international regulatory agencies, QSAR is presently considered as a reliable tool in the risk assessment of chemicals. Traditionally, the QSAR technique has been used to a significant extent in modeling therapeutic activity of drug candidates and toxicity profile of chemicals along with the prediction of various physicochemical parameters. Successful application of QSAR also includes the design and development of agrochemicals. Apart from these, QSAR is presently employed in various emerging fields [6]. Modeling of different property endpoints in material informatics can be cited in this regard. QSAR is being successfully used for modeling response data derived from cosmetics, peptides, phytochemicals, catalysts, polymers, ceramics, novel chemicals such as nanoparticles (NPs) (such as fullerenes, metal oxide nanoparticles, and carbon nanotubes), ionic liquids, and supercritical carbon dioxide. It is to be noted that the mentioned areas do not give an exhaustive list for the newer application areas of QSAR technique. The formalism of QSAR can also be used to address newer fields in chemical research based on the intuitive experimental design and the purpose involved.

4.2 Newer Methods

4.2.1 HQSAR

4.2.1.1 Perception of HQSAR

Hologram QSAR (HQSAR) is a 2D fragment-based newer technique dependent on the concept of employing molecular substructures expressed in a binary pattern (i.e., 2D fingerprints) as descriptors in QSAR models [1]. HQSAR does not require any physicochemical descriptors or 3D structures to generate the structure–activity model. Thus, 2D structures and biological activity are employed as inputs, and the structures are converted to all possible linear, branched, and overlapping fragments. The generated fragments are assigned to integer values using a cyclic redundancy check algorithm. These integer values are used to make an integer array of fixed length. These arrays are considered as molecular hologram, and space occupancies of the molecular holograms are utilized as descriptors. Finally, partial least squares (PLS) regression is used to construct the model which is validated by the leave-one-out method. The obtained model equation should be like the following:

$$A_i = C + \sum_{i=1}^{L} X_{il} C_{il} \qquad (4.1)$$

where A_i is the activity of compound i, X_{il} is the hologram occupancy value at position i or bin l, C is a constant, C_{il} is the coefficient for the corresponding bin from the PLS run, and L is the hologram length.

4.2.1.2 Methodology

The HQSAR methodology consists of three fundamental steps [2]:

(i)   Generation of substructural fragments for each of the training set molecules,
(ii) Representation of the structural fragments in the form of holograms,
(iii) Thereafter, correlation of the molecular holograms with the activity data of the training set molecules employing the PLS tool to generate a HQSAR model.

A graphical depiction of creation of molecular holograms and a HQSAR model is illustrated in Fig. 4.1.

4.2.1.3 HQSAR Parameters

The performance of HQSAR models can be influenced by a number of parameters considering hologram generation.

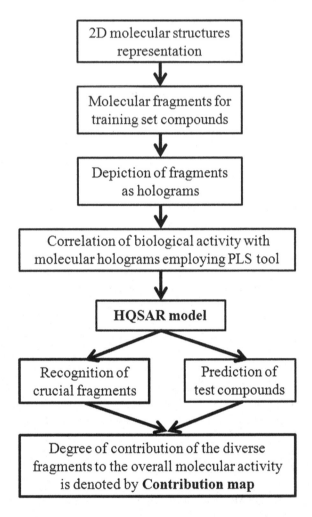

Fig. 4.1 A schematic diagram of HQSAR methodology

- **Hologram length**: The hologram length is a user-defined parameter which controls the number of bins in the hologram fingerprint.
- **Fragment size**: Fragment size controls the minimum and maximum lengths of fragments to be included in the hologram fingerprint.
- **Fragment distinction**: HQSAR allows fragments to be distinguished based on various parameters such as atoms, bonds, connections, hydrogens, chirality, donor, and acceptor.

4.2.1.4 Application of HQSAR Models

The major application of the HQSAR study is to explore individual atomic contribution to molecular bioactivity with a visual display of active centers in the compounds. The HQSAR has been successfully applied in various stages of drug discovery process in recent times.

- **Versatile tool in drug design**: Along with the prediction of potency and affinity of new compounds, HQSAR models are capable of providing constructive insights into the relationships between structural fragments and biological activity.
- **A flexible tool for virtual screening (VS)**: The basic approach of HQSAR can be delicately applied in the VS strategies for the identification of hits. In case of large data sets generated by combinatorial chemistry and high-throughput screening (HTS) techniques, there are a variety of applications of HQSAR.
- **Pharmacokinetic/pharmacodynamic studies and ADME prediction**: The HQSAR patterns of substructural fragments could also be helpful in pharmacokinetic studies comparing conventional mechanism-based pharmacodynamic modeling. The recognized substructural patterns for a particular group of compounds can be utilized as ADME filters in design of chemical library and VS.

4.2.1.5 Advantages of HQSAR

The technique offers the following advantages:

- HQSAR provides a precise prediction of the activity of untested molecules.
- The technique eliminates the need for generation of 3D structures, putative binding conformations, and molecular alignments.
- The approach provides a visual display of the active centers in compounds indicating the fragments contributing maximally to the activity profile of the compounds.

4.2.2 G-QSAR

4.2.2.1 Idea Behind Group-based QSAR (G-QSAR)

G-QSAR [3] is a fragment-based QSAR tool which is capable of establishing a correlation of chemical group variation at diverse molecular sites of interest with the consequent biological activity. The G-QSAR technique forms a mathematical equation between the activity and descriptors computed for a variety of molecular fragments of interest using specific fragmentation rules. The novelty of the G-QSAR approach lies in the interpretation of the indispensable requisites of the different substituents by suggesting not only the imperative descriptors but also reflecting the site where one has to optimize for the design of new active compounds.

4.2.2.2 G-QSAR Methodology

The G-QSAR techniques can be accurately described in three steps as discussed below:

(a) *Fragmentation of compounds*: The first and foremost step of the G-QSAR technique is the fragmentation of compounds under study. In case of a set of congeneric molecules, the number of fragmentation sites depends on the substituents present in the core scaffold. For a non-congeneric set of molecules, fragmentation of a set of molecules is performed using a predefined set of rules.

(b) *Descriptor computation of individual fragment*: The next step is the computation of descriptors for each fragment of a given molecule in the following manner: At first, 2D/3D descriptors are calculated for fragments present in individual molecules in the data set, and secondly, along with the 2D/3D descriptors, cross-interaction terms between diverse fragments are also computed.

(c) *Construction of G-QSAR model*: The final step is the selection of the best possible set of descriptors from the entire pool of descriptors to create the QSAR model. For the selection of an optimal subset of descriptors, various variable selection methods (stepwise forward, stepwise forward-backward, stepwise backward, simulated annealing method, genetic algorithm, etc.) can be employed. Tools such as multiple linear regression (MLR), principal component regression (PCR), PLS, k-nearest neighbor, and neural networks, are used to develop the final QSAR model. Figure 4.2 represents a complete schematic diagram of the G-QSAR methodology.

4.2.2.3 Advantage of G-QSAR

The major advantage of the G-QSAR technique is that it considers the substituent interactions as fragment specific descriptors to account for the fragment interactions in the QSAR model. Other 2D-QSAR approaches are capable of suggesting only important fragments, whereas the G-QSAR approach can reflect the crucial descriptors along with the site where it has to be optimized for the design of new molecules.

4.2.2.4 Application of G-QSAR Model

The G-QSAR technique offers identification of required structural modifications at specific substitution sites and also provides a predictive model for the future prediction of new chemical entities (NCE). The site specific precisions along with the interpretation of fragments are determined from the G-QSAR model.

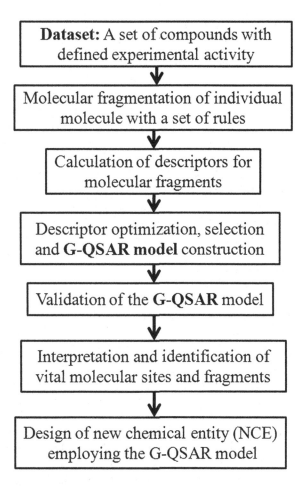

Fig. 4.2 Fundamental steps of G-QSAR methodology

- **Drug design employing requisite fragments**: G-QSAR models are capable of generating information about the fragments which contribute significantly to the variation in activity. Analysis of the fragments with the reference chemical structure can provide essential information for the new drug design.
- **Scaffold hopping and lead optimization**: G-QSAR is independent of 3D conformations and alignment of the molecules and can be employed tacitly for scaffold hopping and lead optimization by employing descriptors of selected fragment(s) of active molecules.
- **Solution of '*inverse QSAR*' problem**: G-QSAR addresses 'inverse QSAR' problem which offers a systematic method to design molecules that satisfy QSAR necessities and thereby design active molecules.
- **Prediction of activity response through mathematical equation**: The G-QSAR method can be efficiently applied for the prediction of databases from different classes of activities.

4.2.3 MIA-QSAR

4.2.3.1 Concept of MIA-QSAR

MIA has one of the most widespread practical applications in the fields of scientific imaging. The MIA is a type of multivariate regression method that is based on data sets obtained from 2D images. Freitas et al. [4] have constructed a simplified QSAR method based on 2D images of congeneric series of compounds. In MIA-QSAR, 2D images are employed for the generation of pixels for individual compounds of the studied data set. Thereafter, the computed pixels of individual images are considered as descriptors which are correlated with the respective biological response for the generation of QSAR models. The MIA-QSAR is capable of testing a good number of compounds with diverse substituents in order to verify the variation of activity for the specific group of compounds. This is a direct visual tool to predict biological activity in a quantitative way for a series of molecules with congenericity.

4.2.3.2 Methodology of MIA-QSAR

(a) *Computation of descriptors*: The 2D image generated information (here, pixels) is used as descriptors for the MIA-QSAR. As described earlier, in case of MIA-QSAR, multivariate images are employed for computation of descriptors. In case of multivariate images, each image is a 3D array with height × width × wavelength dimensions. For instance, the most common type of multivariate images presents a color image where wavelengths corresponding to red, green, and blue lights are measured, respectively. Thus, dimensions of the 3D array of these types of images are expressed as height × width × 3 (where 3 represents red, green, and blue wavelengths). After generating binaries of each image, they are superimposed to create a tensor. The generated tensor is unfolded in order to use two-way analysis. Therefore, the generated 2D matrix for the whole data set can be utilized as the total pool of descriptors.

(b) *Model generation*: The descriptor matrix formed by the pixels is consequently decomposed into a score vector s_1 and a weight vector w_1. The score vector is determined to have the property of maximum covariance with the dependent variable y. The score vectors then replace the original variables as regressors. Due to the implication of pixels as descriptors, the problem of collinearity and noise is a serious concern for MIA-QSAR. Methods such as principal component analysis and PLS regression can be used to avoid the use of collinear descriptors as these methods generate new orthogonal descriptors resulting in better robust and predictive models. Then, the final descriptor matrix is correlated with the biological response value by means of any suitable chemometric tools. The generated equation from the training set of compounds should be employed for the prediction of test set and subsequently utilized for the recognition of critical structural requisites for the enhanced activity. The methodology of MIA-QSAR is depicted in a schematic diagram in Fig. 4.3.

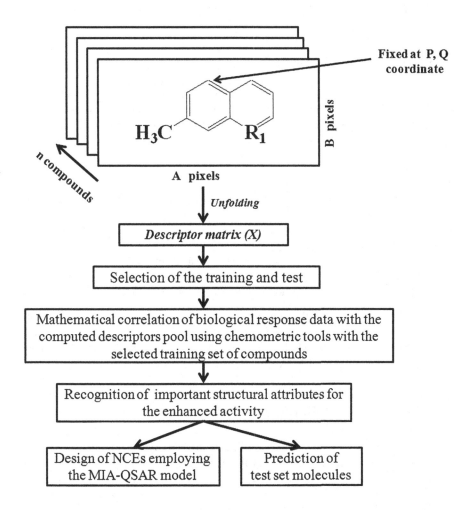

Fig. 4.3 Flowchart of MIA-QSAR formalism

4.2.3.3 Advantages of MIA-QSAR

- MIA-QSAR is a simple and fast 2D-QSAR technique.
- It is possible to design molecules with diverse substituents.
- It is capable of predicting any modeled response in a direct or visual way especially for a congeneric series of molecules.
- The approach does not require 3D alignment as well as conformational analysis.

4.2.3.4 Drawbacks of MIA-QSAR

As the descriptor calculation in MIA-QSAR is solely dependent on the 2D images of individual molecules, therefore, drawing and representation of molecular structure has a decisive role to play in the development of statistically robust and interpretable models. Factors which may influence the descriptor calculation are illustrated below.

- Font size in the drawn structures
- Font type used in the drawn structures (Arial, Times New Roman, Comic Sans MS, etc.)
- Representation of substituents (for example, an ethoxy group can presented in the form of OEt or OC_2H_5)
- Image saving format [format: images can be saved in different standard format such as joint photographic experts group (JPEG), tagged image file format (TIFF), bitmap (BMP), and portable network graphics (PNG)].

As a consequence, with the change of font type and size, depiction of substituents as well as image saving format, pixel numbers will vary for not only each substituent but also for the whole molecule. The mentioned problems may largely influence the consistency of MIA-QSAR.

4.2.3.5 Application of MIA-QSAR

- MIA-QSAR can be successfully applied in the activity prediction.
- Recognition of crucial structural attributes for activity profile of a specific class of compounds is possible which leads to further NCE design and creation of database for future.
- The created databases may be employed for future scaffold hopping and lead optimization without applying conformational analysis or any alignment techniques.

4.2.4 Binary QSAR

4.2.4.1 Concept of Binary QSAR

The beginning of combinatorial chemistry for the development of large chemical libraries constrained scientist to discover fast approaches for assaying millions of chemicals at a go. For this purpose, the HTS is an ideal technique. There is always a chance of error prone results in this method, and on the other hand, the conventional QSAR approaches require more homogenous compounds with a continuous activity data. To address the existing problems in traditional QSAR techniques and to handle a large number of binary data from HTS, the 'binary QSAR' method was implemented, which can handle data from HTS [7]. The method considers binary activity measurements in the form of actives or inactives and computed molecular descriptor vectors as input. A Bayesian inference technique is used to predict whether or not a new compound will be active or inactive.

4.2.4.2 Methodology of Binary QSAR

In a simplified explanation, binary QSAR correlates compound structures employing molecular descriptors, with a 'binary' expression of activity (i.e., 1 = active and 0 = inactive), and computes a probability distribution for active and inactive compounds in the training set. This function can then be employed to predict active compounds for a given target in a test set. The methodology of the binary QSAR is illustrated below:

(i)   Representation of individual compound structure.
(ii) Computation of molecular descriptors and construction of a matrix table consisting of descriptors and biological response of corresponding compounds.
(iii) Then, the original molecular descriptors are transformed to a decorrelated and normalized set of descriptors. The desired probability density is then approximated by applying Bayes' theorem and assuming that the transformed descriptors are mutually independent.
(iv) Finally, construction of binary QSAR model employing any suitable chemometric tools.

A simplified flowchart of binary QSAR is presented in Fig. 4.4.

4.2.4.3 Advantage of Binary QSAR

The major advantage of this method is that it is useful for prioritizing compounds for HTS, for construction of combinatorial libraries, and for screening and synthesizing virtual libraries.

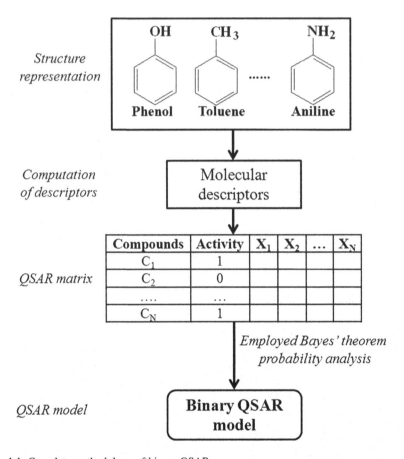

Structure representation

Computation of descriptors

QSAR matrix

QSAR model

Fig. 4.4 Complete methodology of binary QSAR

4.2.4.4 Drawbacks of Binary QSAR

Binary QSAR assigns a probability to a compound to be active in a particular test setting, but it cannot predict specific modifications of lead compounds to enhance their activity. Therefore, binary QSAR is not an alternative to classical QSAR analysis. Another drawback is the complexity of interpreting the importance of descriptors in the constructed model.

4.2.5 Miscellaneous Methods

Comparatively new, emerging, and less employed QSAR techniques are listed in the Table 4.1 for a better understanding of these approaches for future research [5].

Table 4.1 Miscellaneous newer QSAR methods

Methods	Concept/explanation
Ensemble QSAR (eQSAR)	It is a novel QSAR technique which addresses the significance of low-energy conformers in QSAR analysis defined as ensemble QSAR (eQSAR). The term "ensemble" depicts the efforts at imitating the conformational space of the ligand by using a finite set of low-energy conformations. The originality of this method is that the biological response is modeled as a function of physicochemical description initiating from an ensemble of low-energy (active) conformers, rather than as a property generated from the single lowest energy gas phase conformer. The methodology is capable of predicting whether a particular structural modification would improve or hinder drug binding
Fragment-based QSAR (FB-QSAR)	The FB-QSAR is an extension of both the Free–Wilson QSAR and the classical 2D-QSAR. The novelty of the new method is that the compound is partitioned into several fragments based on their substitutions. The response of the molecules is correlated with the physicochemical properties of the molecular fragments through two sets of coefficients (physicochemical properties and weight factors of the molecular fragments) in the linear free-energy equations
Fragment similarity-based QSAR (FS-QSAR)	The FS-QSAR was developed to determine the major restriction of the original Free-Wilson method by introducing the fragment similarity concept in the linear regression equation. In case of FS-QSAR, the fragment similarity calculation was carried out by the similarity. The method used the lowest or highest eigenvalues computed from BCUT-matrices, which contained partial charges of individual atoms and their atomic connection information in each individual fragments
LQTA–QSAR	*Laboratório de Quimiometria Teórica e Aplicada* (LQTA) investigates the main features of CoMFA and 4D-QSAR paradigms where conformational flexibility is commonly studied. This approach creates conformational ensemble profile (CEP) for each compound instead of only one conformation. After that, the molecular dynamics (MD) trajectories and topology information retrieved from the GROMACS free package are used for the calculation of 3D descriptors. The GROMACS computes the intermolecular interaction energies at each grid point considering probes and all aligned conformations resulting from MD simulations. These interaction energies are the independent variables or descriptors which are employed in the QSAR analysis. The LQTA-QSAR is an open access tool for the scientific community at http://lqta.iqm.unicamp.br

(continued)

Table 4.1 (continued)

Methods	Concept/explanation
Receptor-dependent (RD) 4D-QSAR	Receptor-dependent 4D-QSAR is a relatively new approach in QSAR where experimental techniques like X-ray crystallography, NMR spectroscopy and homology modeling are utilized to find out 3D structures of macromolecules. The 3D structure is determined and the binding site for the ligand is predicted which permits to know the binding and alignment modes of ligands. The basic aim of the RD 4D-QSAR study is to map the ligand–receptor interaction mode
Receptor-independent (RI) 4D-QSAR	The RI 4D-QSAR is employed either to find the pharmacophoric features of the ligand or to figure out the projected changes in ligand structure. The aim of RI 4D-QSAR is to attain maximum structural information from the developed model. The advantage of RI is that it will design and construct pharmacophoric features of the substituents and design and map rational base for substituent placement on the scaffold which can be employed as an initial filter in virtual screening
SOM 4D-QSAR	The self-organizing map (SOM) is a machine learning tool employed to classify the data according to the similarity. It is a common type of artificial neural network (ANN) which is frequently used in the QSAR due to its precision and simple interpretation. This method is applicable for experiments where an active bound conformation is searched taking into account conformation flexibility
5D-QSAR (QUASAR)	It considers the multiple expression of ligand topology to study conformation, isosteriomer, and protonation, while orientation is the new dimension added to 4D-QSAR as it can be represented in multiple induced fit and referred as the 5D-QSAR analysis
6D-QSAR	6D-QSAR study considers the solvation function in QSAR analysis which is an expansion of the QUASAR (5D-QSAR) where employing the consideration of simulations for different solvation models
7D-QSAR	One more dimension has been added to the 6D-QSAR to introduce another higher dimension QSAR (7D-QSAR). The 7D-QSAR analysis comprises real receptor or target-based receptor model data

4.3 Future Scope

4.3.1 What to Expect in the Coming Days

Though QSAR is basically a ligand-based statistical approach, a combination of QSAR with receptor-based approaches has demonstrated useful applications and success for optimization of drug candidates. QSAR is also useful for pharmaco-kinetic data modeling. QSAR/QSPR has emerged as an alternative method for risk assessment of chemicals in the context of environmental safety. Starting with the classical Hansch and Free–Wilson approaches, QSAR/QSPR has gradually evolved with time through refinement of approaches, use of newer descriptors, application of diverse chemometric tools, employment of rigorous validation tests, and integration with receptor structure information.

4.3.2 Newer Application Areas of QSAR/QSPR

Apart from its use in ligand optimization in the context of drug discovery and predictive risk assessment in ecotoxicology, there are several new emerging fields [6] in which QSAR/QSPR is finding its application. Herein, we list some other areas which will find potential applications of QSAR/QSPR in the coming days.

4.3.2.1 QSAR of Nanoparticles

NPs have found a wide range of applications in industrial sectors and different fields of human life. QSAR modeling might be applicable for the comprehensive risk exposure and assessment of NPs at the early stage of their development. A new term 'nano-QSAR' has recently been coined. Efforts should be made to develop new descriptors and methodologies for developing QSAR models for this special class of chemicals.

4.3.2.2 QSAR of Mixture Toxicity

In the environment, different chemicals remain in a mixture form, which may behave in a different way from the pure chemicals due to the interactions with and effects of other chemicals. It will be an important and interesting research area to develop QSAR models for predicting toxicity of mixtures. The QSAR modeling of mixtures requires the use of appropriate descriptors. Efforts are to be directed to the development of new descriptors and the improvement of existing QSAR approaches for mixtures.

4.3.2.3 QSAR of Peptides

Antimicrobial peptides have recently drawn significant attention as an alternative class of antimicrobial therapeutics. However, their structure–activity relationships (SAR) are not well understood largely because of substantial diversity in their structures and their non-specific mechanism of action. There is possibility of application of QSAR in further understanding their SAR.

4.3.2.4 QSAR of Cosmetics

QSAR modeling may emerge as one of the leading alternatives of animal studies for testing of safety and risk assessment of cosmetics. The various toxicological end-points for the development of QSAR models relevant to cosmetics are as follows: acute toxicity, skin irritation and corrosion, skin sensitization, dermal absorption, mutagenicity/genotoxicity, carcinogenicity, reproductive toxicity, etc.

4.3.2.5 QSAR of Ionic Liquids

Ionic liquids, a relatively new class of chemicals promoted as green solvents, have diverse application in synthetic chemistry, electrochemistry, analytical chemistry, separation and extraction, and other engineering and biological applications. The experimental property and/or toxicity data have been reported only for a small fraction of them. QSPR/QSAR of ionic liquids may help to design suitable combination of cations and anions leading to 'greener' solvents with desirable properties and reduced toxicities.

4.3.2.6 Material Informatics

The theory of QSPR modeling may be applied to different areas of material sciences such as rubber chemistry and chemistry of fullerenes, catalysis, and biomaterials.

4.3.2.7 Interspecies Toxicity Modeling

Interspecies toxicity correlations provide a tool for estimating a contaminant's sensitivity with known levels of uncertainty for a diversity of different species. This approach can be applied for reduction of animal testing by gathering and extrapolating information from tested to untested species, as well as from tested to untested chemicals.

4.3.2.8 QSAR of Phytochemicals

There is an increasing use of novel plant products and chemical libraries based on phytochemicals in drug discovery programs. However, only a limited number of in silico models have been reported so far in the literature based on phytochemicals. There is ample scope of application of QSAR modeling in the field of phytochemicals for exploring newer drug candidates.

4.4 Conclusion

The journey of the QSAR/QSPR study has traversed a long path starting with the classical approaches of Hammett, Taft, Hansch, and Free–Wilson. The gradual evolution of the formalism has included the development of new dimensional concepts, new descriptors, various chemometric tools as well as statistical validation criteria. The findings of the QSAR/QSPR study are also enriched by its combined use with various structure-based approaches. The refinement also involves various newer QSAR techniques intending to provide more prompt and accurate information on the chemical attributes. QSAR has now evolved as a distinct scientific discipline on its own merit. Along with the application in the optimization of ligands in drug discovery and predictive risk assessment, the QSAR method has been found to be useful in various other emerging research areas. The seminal guidelines prescribed by the OECD have enabled the users to develop robust and predictive QSAR models. Further introspection of biological, toxicological as well as physicochemical endpoints can enhance the application opportunities of the QSAR idealism.

References

1. Lowis DR (1997) HQSAR a new, highly predictive QSAR technique. Tripos Tech Notes 1:1–10
2. Doddareddy MR, Lee YJ, Cho YS, Choi KI, Koh HY, Pae AN (2004) Hologram quantitative structure activity relationship studies on 5-HT6 antagonists. Bioorg Med Chem 12 (14):3815–3824
3. Ajmani S, Jadhav K, Kulkarni SA (2009) Group-based QSAR (G-QSAR): mitigating interpretation challenges in QSAR. QSAR Comb Sci 28(1):36–51
4. Freitas MP, Brown SD, Martins JA (2005) MIA-QSAR: a simple 2D image-based approach for quantitative structure–activity relationship analysis. J Mol Struct 738:149–154
5. Damale MG, Harke SN, Khan FAK, Shinde DB, Sangshetti JN (2014) Recent advances in multidimensional QSAR (4D-6D): a critical review. Mini-Rev Med Chem 14:35–55
6. Cherkasov A, Muratov EN, Fourches D, Varnek A, Baskin II, Cronin M, Dearden J, Gramatica P, Martin YC, Todeschini R, Consonni V, Kuz'Min VE, Cramer R, Benigni R, Yang C, Rathman J, Terfloth L, Gasteiger J, Richard A, Tropsha A (2014) QSAR modeling: where have you been? Where are you going to? J Med Chem 57:4977–5010
7. Gao H, Williams C, Labute P, Bajorath J (1999) Binary quantitative structure-activity relationship (QSAR) analysis of estrogen receptor ligands. J Chem Inf Comput Sci 39:164–168

Printed in the United States
By Bookmasters